An Approach to Improving Decision Making in Wetland Restoration and Creation

An Approach to Improving Decision Making in Wetland Restoration and Creation

Mary E. Kentula[1]
Robert P. Brooks[2]
Stephanie E. Gwin[3]
Cindy C. Holland[3]
Arthur D. Sherman[3]
Jean C. Sifneos[3]

Edited by

Ann J. Hairston[3]

[1]U.S. Environmental Protection Agency
Environmental Research Laboratory
Corvallis, OR

[2]The Pennsylvania State University
Forest Resources Laboratory
University Park, PA

[3]ManTech Environmental Technology, Inc.
USEPA Environmental Research Laboratory
Corvallis, OR

Project Officer

Mary E. Kentula

Wetland's Research Program
U.S. Environmental Protection Agency
Environmental Research Laboratory
Corvallis, OR

C. K. SMOLEY, INC.

Library of Congress Cataloging-in-Publication Data

Catalog record is available from the Library of Congress.

ISBN 0-87371-937-9

© 1993 by C. K. SMOLEY

Direct all inquiries to CRC Press Inc.
2000 Corporate Blvd N.W.
Boca Raton, FL 33431

PRINTED IN THE UNITED STATES OF AMERICA
1 2 3 4 5 6 7 8 9 0
Printed on acid-free paper

PREFACE

As wetland losses continue and restoration and creation efforts increase, long-term research data become essential to understanding the impacts of our regulatory decisions. AN APPROACH TO IMPROVING DECISION MAKING IN WETLAND RESTORATION AND CREATION is the culmination of five years of research primarily in Connecticut, Florida, and Oregon. This research compares populations of natural and created wetlands to determine whether restored wetlands successfully replace wetlands lost to development and other pressures. The type of information synthesized in this document can be used by resource managers in determining strategies for mitigation of wetland losses. In addition, this approach addresses management concerns such as site selection for future restoration projects, assessment of the level of attainable function for restored wetlands, and how to evaluate when the desired level of function has been achieved.

Although primarily designed to meet the needs of the EPA regions and the Office of Water, this is a useful document that will undoubtedly be read with varying expectations by a wide audience. There are ideas for all readers. The approach offered makes a significant contribution to the scientific information base for decisions on wetland restoration and creation. This is a synthesis document and much additional information can be found by consulting the original research results. This approach is not intended to define EPA policy. And, although we do not endorse any one approach, we can certainly endorse the main theme of this document, "we must learn from what we have done and use that information to improve future resource management".

Wetlands Division
Office of Wetlands, Oceans, and Watersheds

DISCLAIMER

The research described in this document has been funded by the United States Environmental Protection Agency under Contract #68-C8-0006 to Man-Tech Environmental Technology, Inc. and Contract #68-C0-0021 to Technical Resources, Inc. Mention of trade names or commercial products does not constitute endorsement or recommendation for use.

This document should be cited as:

    Kentula, M.E., R.P. Brooks, S.E. Gwin, C.C. Holland, A.D. Sherman, and J.C. Sifneos. 1992. An Approach to Improving Decision Making in Wetland Restoration and Creation. Edited by A.J. Hairston. U.S. Environmental Protection Agency, Environmental Research Laboratory, Corvallis, OR.

LETTER

To the Reader,

We have kept you in mind throughout the preparation of DECISION MAK-ING. In particular, we have attempted to fill a gap in the information available on wetland restoration and creation by addressing how to improve future decisions by evaluating the results of past decisions. We are aware of the varied needs and interests of our audience and feel a responsibility to meet the expectations those needs and interests bring to a reading of our work. Reflecting this diversity, the reviews of our draft manuscript indicated that many readers had a "favorite" or "most useful" chapter and that there was little agreement as to which of the chapters it was! We learned that DECISION MAKING could be read and used in a number of ways we had not anticipated.

We recommend that you review the Table of Contents before beginning to read the book and use the chapter and section titles to select the parts that will be of most interest to you. Each chapter was written both to fit into the framework of the book and to present a single concept. DECISION MAKING does not have to be read in order, from front to back. The only exception is Chapter 1, which should be read first because it defines the terms we use. After that, if you are interested in monitoring projects, go to Chapter 4, OR jump to Chapter 6 to read about evaluating design, OR go on to Chapter 2. Whichever route you choose, we hope that you will find much of value.

The Authors

EXECUTIVE SUMMARY

AN APPROACH TO IMPROVING DECISION MAKING IN WETLAND RESTORATION AND CREATION

The U.S. Environmental Protection Agency's Wetlands Research Program (WRP) has developed an approach to improving decision making in wetland restoration and creation projects. The WRP Approach uses data from a monitoring program, including both natural wetlands and those restored and created, to develop performance criteria, track the development of projects, and suggest improvements in the design of future projects. For the past five years, scientists in association with the WRP have been developing the Approach by comparing the characteristics of mitigation projects and natural wetlands to test methods for data collection, and to evaluate project design and compliance with permit conditions. Many of the same methods were used in studies in Connecticut; Tampa, Florida; Portland, Oregon; and Seaside, Oregon, so that the techniques could be evaluated, the findings from all studies compared and the results used to refine the Approach.

The projects studied were typically less than or equal to five years old, and the majority were what is probably the most common freshwater mitigation project nationally—a pond with a fringe of emergent marsh. We chose this type of project because they were abundant, comprising a major proportion of the compensatory mitigation projects required nationally under Section 404 of the Clean Water Act. Because the Approach was developed in freshwater systems, the monitoring techniques and examples presented will transfer most readily to freshwater nontidal wetlands. However, application of the Approach is not limited to either mitigation projects or freshwater nontidal wetlands; it is determined by the needs of the agency or organization involved, and ultimately, the status of the wetland resource.

The Approach is based on the assumption that natural wetlands in a region can be used as models to define the standards for restoration and creation projects. Comparison of wetland projects with natural wetlands located in a similar land use setting and, therefore exposed to similar ecological conditions, is important to ensure that what is "expected" of a project is within the bounds of possible performance. Major recommendations are:

- Use information in project files to guide decision making.

- Target areas at greatest risk.

- Base the level of effort used in monitoring on information needs.

- Consider the landscape setting of the wetlands when defining the populations to be compared.

- Use the characteristics of natural wetlands and wetland projects to define the standard.

- Make the process of setting performance criteria and defining design guidelines iterative.

CHAPTER 1 presents an overview of the WRP Approach. This includes discussion of the above recommendations and the major analytical tool of the Approach, the performance curve. The performance curve documents the development of the ecological function of wetland projects over time relative to similar natural wetlands. We envision that a set of performance curves will be produced for each function or indicator measured. What is measured is determined by the goals of the resource management program and the specific projects. Management questions that can be answered using this strategy include:

- What level of function is achievable for natural wetlands and wetland projects in a particular land use setting?

- Do the projects achieve the level of function of natural wetlands in similar settings?

- How long does it take for projects to achieve the desired level of function?

In a mitigation context, answers to such questions would allow managers to identify which permits should 1) be most critically reviewed because of low probability of successful mitigation; 2) require the most comprehensive checks on design and implementation of the mitigation project because of uncertain probability of success; and 3) require minimal checks on design and implementation of the mitigation project because of high probability of success.

CHAPTER 2 details how to use the information from project files in decision making. If the data are to be of use in protecting the wetland resource, they must be updated, compiled, analyzed, and reported. For example, analysis of previous trends in permitting can reveal locales and wetland types subject to the most intense permitting activity. With knowledge of such trends, permitting agencies can take action to avoid potential losses in wetland number, type, function, and area.

EXECUTIVE SUMMARY

CHAPTER 3 describes a method for sampling populations of projects and natural wetlands to select sites for study. Targeting efforts to areas where the resource is, or is predicted to be, at risk is also discussed. The Approach describes how to 1) define the population of projects to be sampled; 2) use the location of the sites to define the boundaries of the study area within the area at risk; 3) use the characteristics of the population of projects to define the population of natural wetlands to be sampled; 4) randomly select sites from the populations of projects and natural wetlands; and 5) finalize the list of sites to be sampled by verifying that the sites exist, are accessible, and belong to the populations defined.

CHAPTER 4 presents a post-construction monitoring strategy that recommends three levels of sampling depending on the age and goals of the project: documentation of as-built conditions; routine assessments; and comprehensive assessments. For each of the variables suggested, a brief rationale relating it to wetland function is provided. Finally, components integral to a post-construction monitoring plan, such as maintaining data quality, timing of sample collection, and controlling damage to the site, are discussed.

A major obstacle to long-term monitoring and, therefore, implementation of the WRP Approach is cost. The special insert, **Volunteers and Natural Resource Monitoring**, that follows Chapter 4, presents a possible low-cost, high-profit solution. WRP scientists worked closely with Neal Maine, an award winning science education specialist, to train and coordinate a group of citizen volunteers to assist in the study of the Trail's End mitigation project near Seaside, Oregon. The study makes the connection between scientific research and public education which is so often ignored. The research profited because more types of data were collected over a longer period of time than would have been possible otherwise. The community profited because volunteers who were teachers taught their students to collect data on local wetlands using the techniques learned from the study.

In **CHAPTER 5** four different types of graphs are suggested for representing monitoring data: performance curves, summary or descriptive graphs, time series graphs, and characterization curves. Statistical methods are outlined for data manipulation that will enable resource managers to organize incoming data to track the progress of projects, and to develop criteria for the evaluation of future projects. Graphical displays are used to illustrate how to evaluate projects and set performance criteria.

CHAPTER 6 illustrates how data collected from local natural wetlands can, and should, be used to improve the design of projects. This chapter details design features, including type of wetland; slopes of banks; amount of area; and appropriate hydrology, vegetation, and soils/substrates. Development of a planting list of species appropriate to a specific wetland type and locale is presented as an example of how to tailor project design to meet local needs.

We have developed the WRP Approach to help anyone working to protect the wetland resource use past data from restoration or creation projects as a management tool to improve decision making and, thereby, the ability to restore and create wetlands in the future. Our philosophy is that by considering the surrounding land use, comparable natural wetlands, and similar projects, you can design wetland projects with a better chance of long-term success. Determining the effects of different land uses on wetland function will be a major theme of our upcoming research. Such information will be important to both the protection of the wetland resource and the success of restoration and creation projects. With knowledge of the effects of surrounding land uses, appropriate management strategies can be employed to protect key wetlands. In addition, knowing how present and projected development of an area will affect wetland function can influence decisions on how to prioritize restoration sites for maximum ecological benefits.

Fundamentally, as we plan and implement new studies we will continue to treat existing projects as experiments in progress and to promote the idea that we all must "...learn by going where we need to go..." (Roethke 1961).

CONTENTS

PREFACE v

DISCLAIMER vi

TO THE READER vii

EXECUTIVE SUMMARY ix

LIST OF TABLES xvi

LIST OF FIGURES xvii

BACKGROUND AND ACKNOWLEDGEMENTS xxi
    AN UPDATE ON THE STATUS OF THE SCIENCE xxii
    THE RESEARCH STRATEGY USED BY THE WRP xxii
    ACKNOWLEDGEMENTS xxiii

CHAPTER 1 1
    TERMS USED 2
    THE WRP APPROACH AND ITS APPLICATIONS 2
    KEY CONCEPTS 3
        Populations 3
        Setting 5
        Performance Curves 5
        Indicators 7
    SUMMARY 8

CHAPTER 2 11
    MINIMUM INFORMATION NEEDED 12
    FEATURES OF EPA's PERMIT TRACKING SYSTEM 13
    INCORPORATING ADDITIONAL INFORMATION 15
    REPORTING THE INFORMATION 17
    SUMMARY 19

CHAPTER 3 23
    DECIDING ON A SAMPLING STRATEGY 23
    IDENTIFYING PRIORITY AREAS 24
    SELECTING SITES 26
        Defining the Population of Wetland Projects to Sample 26

An Approach to Improving Decision Making in Wetland Restoration and Creation

CONTENTS

Defining the Boundaries of a Study Area...33
    Taking a regional perspective 33
    Considering ecological setting..33
Defining and Sampling the Population of Natural Wetlands.............36
Finalizing the List of Projects and Natural Wetlands to be Sampled ..36
SUMMARY 37

CHAPTER 4...43
  DOCUMENTATION OF AS-BUILT CONDITIONS 44
    Rationale 44
    What To Include..52
  ROUTINE ASSESSMENTS..56
    Rationale 56
    What To Include..57
  COMPREHENSIVE ASSESSMENTS..59
    Rationale 59
    What To Include..60
  ASSESSMENT VARIABLES 61
    General Information..61
    Morphometry...61
    Hydrology...63
    Substrate 63
    Vegetation 64
    Fauna...66
    Water Quality 67
    Additional Information...67
  DEVELOPING AN EFFICIENT SAMPLING STRATEGY.............................68
    Data Quality 69
    Where To Collect Samples...70
    How Many Samples To Collect..70
    When To Collect Samples..71
    Controlling Damage To The Site 71
  SUMMARY 71

VOLUNTEERS AND NATURAL RESOURCE MONITORING 73

CHAPTER 5...87
  SUGGESTED WAYS TO REPRESENT THE DATA COLLECTED 87
    Performance Curves...88

Contents

Summary or Descriptive Graphs..92
Time Series Graphs 93
Characterization Curves... 93
TECHNIQUES FOR DETERMINING DIFFERENCES IN SAMPLES 96
EVALUATING PROJECTS AND SETTING PERFORMANCE CRITERIA..........................98
An Extension of the Example 105
Example of How to Use Time Series Graphs 105
Example of How to Use Characterization Curves 107
SUMMARY 108

CHAPTER 6..111
WETLAND TYPE..111
Determine if the Project is Typical of Wetlands in the Region 112
Influence of Bank Slopes on Wetland Type 112
Relationship Between Bank Slopes and Wetland Area 114
Determine how much land will be required 116
Design when adequate land is available...................................117
Design when land area available is limited..............................117
VEGETATION 118
Example from the Oregon Study...119
Example from the Florida Study...120
Guidelines for Revegetation of Wetland Projects............................120
To Plant or Not To Plant? 120
Generating a Planting List 123
What species commonly occur on wetlands in the area?................123
Which species are commercially available?.............................124
Narrow the list of species to generate a planting list 124
OTHER IMPORTANT STRUCTURAL CHARACTERISTICS 127
Hydrology..127
Soils/Substrates 130
SUMMARY 131

REFERENCES 135

LIST OF TABLES

Table I.   Recent publications on wetland restoration and creation xxvii

Table 2-1.   Summary of the Section 404 permit databases compiled
                by EPA's Wetlands Research Program......................................12

Table 2-2.   Minimum categories of data on impacted and
                compensatory (created, enhanced, preserved, or restored)
                wetlands recommended for inclusion in a database and data
                categories found in EPA's Permit Tracking System (PTS)...........14

Table 3-1.   Numbers of freshwater mitigation projects in Portland,
                Oregon, by wetland type and size required in Section 404
                permits issued by the U.S. Army Corps of Engineers and
                the Oregon Division of State Lands from January 1987
                through January 1991 32

Table 4-1.   Rationale and uses of variables measured in as-built,
                routine, and comprehensive assessments of wetland
                projects and natural wetlands...45

Table 4-2.   Methods recommended for measuring variables in as-built,
                routine, and comprehensive assessments of wetland
                projects 48

Table 6-1.   Partial list from which to choose species for planting
                on created/restored wetlands in the Willamette Valley,
                Oregon...126

Table 6-2.   The hydrology planned for created wetlands studied in the
                Portland, Oregon metropolitan area in 1987 128

Table 6-3.   A summary of the findings of recent studies of groups of
                wetland projects...133

LIST OF FIGURES

Figure 1-1.   The steps in the WRP Approach for using quantitative
              information to support decision making.....................................4

Figure 1-2.   Hypothetical performance curve illustrating the
              comparison of natural wetlands and projects (in this case
              restored wetlands) of the same type and similar size in the
              same land use setting relative to a measure of wetland
              function...6

Figure 2-1.   Examples of the query and results screens from the
              Permit Tracking System (PTS)...16

Figure 2-2.   Comparison by state of the percent of the Section 404
              permits requiring compensatory mitigation that specified
              monitoring the project with at least one site visit.....................18

Figure 2-3.   Comparison by state of the net change in area of palustrine
              forested wetlands and palustrine emergent wetlands involved
              in Section 404 permits requiring compensatory mitigation
              over the time period analyzed (see Table 2-1).........................18

Figure 3-1.   Hypothetical performance curves...25

Figure 3-2.   Locations of wetland impacts and creations in Oregon
              that occurred between January 1977 and January 1987..........27

Figure 3-3.   Patterns of Section 404 permitting in California and
              Louisiana... 28

Figure 3-4.   An example of a form that can be used to compile
              information on wetland projects...30

Figure 3-5.   Example of a typical mitigation project sampled in the
              Oregon Study 32

Figure 3-6.   An example of how a U.S. Geological Survey topographic
              map can be used to identify subregions...................................34

LIST OF FIGURES

Figure 3-7.   Example of a completed form that can be used during a
               field reconnaissance to collect information on potential
               study sites 38

Figure 4-1.   Example of a Field Map to document as-built conditions
               of a wetland project 53

Figure 4-2.   Profile of a wetland portraying as-built or current conditions
               based on elevational measurements from Basin Morphometry
               Transects (BMT1 & BMT2). Transects must match with those
               shown on maps with aerial views (See Figure 4-1) 54

Figure 4-3.   Map enlarged from U.S. Geological Survey quadrangle
               showing drainage area, surrounding land-use, and wetland
               location 55

Figure 4-4.   Example of a Field Map to document conditions found
               during a routine assessment as compared to the as-built
               condition. Heavy dark line indicates most recent wetland
               perimeter and separates areas of dominant vegetation types.
               Note change in wetland shape as compared to as-built
               conditions shown on Figure 4-1 58

Figure 4-5.   Field crew members taking elevation measurements along
               a transect 62

Figure 4-6.   Field crew members using a Munsell color chart to determine
               soil hue, value, and chroma 64

Figure 4-7.   Botanist reading a vegetation quadrat 65

Figure 4-8.   Field crew member collecting invertebrates from an
               emergence trap 67

Figure 5-1.   Performance curve generated using the mean percent organic
               matter in the upper five-cm of soil 89

Figure 5-2.   Examples of shapes a performance curve might take 90

LIST OF FIGURES

Figure 5-3.   Performance curves generated using plant diversity data 91

Figure 5-4.   Hypothetical performance curves illustrating four different
              patterns of project development that could be used in making
              management decisions 93

Figure 5-5.   Examples of summary or descriptive graphs............................94

Figure 5-6.   Monthly water levels (cm) for a pair of the created and
              natural wetlands 95

Figure 5-7.   Example of hypothetical characterization curve.......................95

Figure 5-8.   Mean percent cover for created and natural wetlands from
              the Oregon Study plotted versus project age...........................99

Figure 5-9.   Box and whisker plot of cover data for created and natural
              wetlands.. 101

Figure 5-10. Performance curve of plant diversity data 102

Figure 5-11. Bar graph of the percent of species overlap between individual
              created and natural wetlands...103

Figure 5-12. Weighted average scores (Wentworth et al. 1988) for the
              type of vegetation found on individual created (C) and
              natural (N) wetlands 104

Figure 5-13. Example of an emergent marsh in the Connecticut Study.......106

Figure 5-14. Example of a pond with a fringe of emergent vegetation
              from the Florida Study 106

Figure 5-15. Characterization curve of percent organic matter 107

Figure 6-1.   Pictures of typical natural (a), and created (b) wetlands in
              the Oregon1 115

LIST OF FIGURES

Figure 6-2. Topographical profiles for typical natural (a) and created (b) wetlands in the Oregon 116

Figure 6-3. Illustration of how to determine the amount of land needed for creating a wetland given the bank slopes and the depth from the ground surface to the water table.............................118

Figure 6-4. Erosion occurring on steep unvegetated banks at a created wetland sampled in the Oregon Study.....................................119

Figure 6-5. Comparison of the number (a) and the percent cover (b) of species found on created wetlands in the Florida Study 121

BACKGROUND AND ACKNOWLEDGEMENTS

The use of restoration and creation of wetlands to mitigate for permitted losses and to enhance the wetland resource requires an evaluation of the efficacy of the practice. The key question is: Do created and restored wetlands develop the same ecological functions as natural wetlands? In an effort that preceded this document and influenced the research that is described in the following chapters, the Environmental Protection Agency's (EPA) Wetlands Research Program (WRP) assembled a team of experts to compile and document the status of the science on wetland creation and restoration. The resulting publication, **Wetland Creation and Restoration: The Status of the Science** (Kusler and Kentula 1990a), took a national view and built on previous work in the field. This book will build on the major findings of that document which were:

(1)   Practical experience and available information vary by wetland type, ecological function, and region of the country. The most quantitative and best documented information is available for Atlantic coastal wetlands. Fewer projects have been implemented on the Gulf and Pacific coasts and, correspondingly, there is less information. Much less is known about restoring or creating inland wetlands.

(2)   Most wetland restoration and creation projects do not have specified goals, complicating efforts to evaluate "success". Success is often rated on compliance with permit requirements or establishment of vegetation. Such measures, however, do not indicate that a project is functioning properly or that it will persist over time.

(3)   Monitoring of wetland restoration and creation projects has been uncommon. Monitoring of sites and comparisons with natural wetlands over time would provide a variety of information including how projects develop over time and how they compare with natural wetlands in the region (Kusler and Kentula 1990b).

Complementing and including the work presented in **Wetland Creation and Restoration**, the U.S. Fish and Wildlife Service (FWS) is maintaining the Wetland Creation/Restoration database to provide a state-of-the-knowledge resource based on the published literature. A hard copy of the bibliographic material contained in the digital database has also been produced (Schneller-McDonald et al. 1989).

Background and Acknowledgements

An Update on the Status of the Science

Reporting on wetland creation and restoration has burgeoned in recent years. Books, manuals, reports, and journal articles have been published on project design, evaluations of projects, and approaches to monitoring. Table I highlights recent publications.

Interest in restoration ecology has also flourished in the past five years. The Society for Ecological Restoration (SER) was established in 1988 and held its first annual meeting in January 1989 (Hughes and Bonnicksen 1990). As of April 1992, membership in the Society had grown to over 1700. Indicative of the demand for information on ecological restoration, SER is now in the process of establishing a journal, **Restoration Ecology**, to accompany its newsletter and its periodical, **Restoration and Management Notes**.

Three recent compendiums on environmental restoration deserve mention. In **Restoration Ecology: A Synthetic Approach to Ecological Research** (Jordan et al. 1987) some two dozen ecologists discuss the heuristic or intellectual value of ecological restoration. Specifically, they address ecological restoration as a way of raising basic questions and testing fundamental hypotheses about the communities and ecosystems being restored, i.e., as a technique for basic research. **Rehabilitating Damaged Ecosystems** (Cairns 1988) takes a broad, eclectic view. Authors representing diverse fields present case histories from a variety of systems, in addition to discussions of planning procedures and approaches to management. Finally, **Environmental Restoration: Science and Strategies for Restoring the Earth** (Berger 1989) reports the results of the 1988 Restoring the Earth Conference, presenting an overview of the most current techniques and processes for restoration, discussions of current issues, and descriptions of restorations of assorted systems.

The Research Strategy Used by the WRP

This document synthesizes the results of over five years of research by the WRP to illustrate and support an approach to evaluating wetland restoration and creation. The strategic plan for the research (Zedler and Kentula 1986) recommended that existing mitigation projects be treated as "experiments in progress". Implementation of the strategy led to the theme of this document—we must learn from what we have done and use the information to improve future decisions.

Several studies produced the information on which this document is based. These studies are grouped under the two lines of research implemented: examination of the patterns and trends in permitting under Section 404 of the Clean Water Act, and evaluation of freshwater mitigation projects. The reports and papers recounting the results are cited throughout this document.

We analyzed portions of the Section 404 permit records from different regions of the country to determine patterns and trends in permitting activity and to document the cumulative effects of the associated management decisions on the resource. Results from these and similar studies can be used to evaluate wetland management practices, especially the use of compensatory mitigation. We also conducted a number of pilot studies to compare characteristics of mitigation projects and natural wetlands, to test approaches and methods for collecting data, and to evaluate project design and compliance with permit conditions. In particular, data from the pilot studies of freshwater mitigation projects in Portland, Oregon; Tampa, Florida; and Connecticut were used. The experiences resulting from the study of a created wetland in Seaside, Oregon, are reported in the special insert, **Volunteers and Natural Resource Monitoring**. The projects examined in all studies were typically less than or equal to five years old, and the majority were what is probably the most common freshwater mitigation project nationally—a pond with a fringe of emergent marsh. In the Oregon Study a group of 11 created wetlands were compared with a group of 12 natural wetlands. In this case the entire population of mitigation projects existing at the time was sampled. In the Florida Study a group of nine created wetlands were compared to a group of nine natural wetlands. In the Connecticut Study, five Connecticut Department of Transportation mitigation projects were paired with five natural wetlands. In both the Florida and Connecticut studies only a subset of the projects in the area were sampled.

The general framework for each study was similar and was provided by the WRP scientists. It included sampling design, methods, quality assurance procedures, and guidelines for data analysis. This was done so that the framework could be evaluated and the findings from all four studies could be compared. In addition, the principal investigators of each study provided a critique of the framework and introduced new components into their respective studies.

ACKNOWLEDGEMENTS

We appreciate the contributions made by many individuals during the preparation of this document. Personnel from the Wetlands Division of EPA's Office of Wetlands, Oceans, and Watersheds have provided constructive, detailed review comments to the authors and have been extremely supportive of the research necessary to bring this book to completion. We acknowledge their responsiveness and realize that their efforts guarantee the WRP is focusing on issues important to the Agency. We continue to appreciate the personal support of John Meagher, Director, Wetlands Division. In particular, we want to thank Doreen Robb, Wetlands Divisions' liaison to the research program, for her efforts to help us communicate effectively and accurately.

Scientific Contributions to This Document

The research that was synthesized in this document was conducted with the help of many people. In the case of the permit studies, Tina Rohm (formerly with Northrop Services, Inc.) developed a data management system to compile the data. Her system is the model for the Permit Tracking System, which was designed and programmed by Robert Gibson (ManTech Environmental Technology, Inc.) of the WRP. Robert used his creativity in computer programming to simplify and expedite data entry and analysis of the permit information.

Jim Good (Oregon State University), Kathy Kunz (formerly with EPA-Region 10), Michael Rylko (EPA-Region 10), Jane Griffith and Sharon Lockhart (formerly with FWS, Laguna Niguel Field Office), Paul Price (Paul Price Associates, Inc.), and Edwin W. Cake (Gulf Environmental Associates) collected and entered the data from Section 404 permit files. Millicent Quammen (FWS, National Wetlands Research Center, Corpus Christi Field Station) served as the project officer for the studies of permitting in Texas, Arkansas, Louisiana, Mississippi, and Alabama. The permit studies are discussed in Chapter 2.

The Oregon Study was led and conducted by the authors, who are members of the WRP. However, it is important to recognize the contributions of WRP's senior geographer, Brooke Abbruzzese (ManTech Environmental Technology, Inc.). Ms. Abbruzzese played a major role in developing the fundamental approach to site selection. The methods she developed for the Oregon Study (Abbruzzese et al. 1988) became the template for subsequent studies. Moreover, she has continued to contribute by providing advice as we tested and refined the methodology. Site selection is discussed in Chapter 3.

In addition we want to acknowledge the work of several interns from the Geography Department at Oregon State University, Corvallis, Oregon. Jack Davis and Eric Hughes performed the spatial analysis and prepared the associated draft maps that resulted in the Section 404 permit maps of Oregon and Louisiana used in Chapter 3. Tracy Smith developed and prepared the final versions of the Field Maps used in Chapter 4.

Dr. Mark T. Brown (Center for Wetlands, University of Florida, Gainesville) led the Florida Study. He and his staff developed a system for sampling a wetland that minimizes the number of times the site is traversed and, therefore, minimizes the damage to the site (Brown 1991). Dr. Brown also assisted in refining our ideas on site selection. In particular, he developed the Landscape Development Intensity (LDI) index as a way to quantify the landscape setting of a wetland as defined by the land uses in the vicinity of the site (Brown 1991). Use of the LDI is discussed in Chapter 3.

Dr. William A. Niering (Connecticut College, New London) led the Connecticut Study. He and graduate student Sheri R. Confer contributed another dimension to our studies by sampling over time. The Oregon, Florida, and Connecticut Studies sampled the wetlands once during the growing season. In addition, Confer and Niering measured water levels and observed animal use monthly for over a year and characterized plant community composition during two consecutive growing seasons (Confer 1990, Confer and Niering In press). The use of such time series data is discussed in Chapter 5.

Finally, Dr. Milton Weller (Texas A&M University, College Station), in conjunction with Dr. Gerald W. Kaufmann (Loras College, Dubuque, Iowa) and Dr. Paul A. Vohs, Jr. (FWS, Iowa Cooperative Fish & Wildlife Research Unit, Ames, Iowa) expanded our ideas by providing significant information we were not able to collect in the other studies. They repeated their pre-impoundment and early post-impoundment studies done almost 30 years earlier when Drs. Kaufmann and Vohs were graduate students studying with Dr. Weller. These studies documented the changes in the wetlands and the associated waterfowl that resulted from the impoundment in 1961 of Elk Creek, a small creek in Worth County, Iowa (Weller et al. 1991). Information on the development of projects over such a long period of time is extremely rare (Kusler and Kentula 1990b), and, therefore, very valuable.

Technical Contributions to This Document

The authors have benefited greatly from the suggestions of those who reviewed this document. The Wetlands Division, EPA Regions 3, 9, and 10, other agencies, and members of the academic community gave generously of their time to provide comments. Richard Coleman coordinated a review by the U.S. Army Corps of Engineers' Wetlands Research Program at the Waterways Experiment Station; Susan Haseltine, by the FWS Northern Prairie Wildlife Research Station; Lee Ischinger, by the FWS National Ecology Research Center; and Virginia Van Sickle-Burkett, by the FWS National Wetland Research Center. Specifically, we thank the following individuals for their thoughtful reviews: Barbara Bedford (*Cornell University*), Marcia Bollman (*ManTech Environmental Technology, Inc.*), Mary M. Davis (*Army Corps of Engineers*), Kate Dwire (*ManTech Environmental Technology, Inc.*), Paul

BACKGROUND AND ACKNOWLEDGEMENTS

Garrett (*Federal Highway Administration*), Jerry Grau (*Fish and Wildlife Service*), Randy Gray (*Soil Conservation Service*), Susan Haseltine (*Fish and Wildlife Service*), Hal Kantrud (*Fish and Wildlife Service*), Tom Kelsch (*EPA Wetlands Division*), Dennis King (*Maryland Institute for Ecological Economics*), Russ Lea (*North Carolina State University*), Anne Marble (*A.D. Marble & Company*), Geoffery Matthews (*National Oceanic and Atmospheric Administration*), Thomas Minello (*National Oceanic and Atmospheric Administration*), William Niering (*Connecticut College*), Philip North (*EPA Region 10, Alaska Operations*), Philip Oshida (*EPA Region 9*), Barry Payne (*Army Corps of Engineers*), Bruce Pugesek (*Fish and Wildlife Service*), Doreen Robb (*EPA Wetlands Division*), Charles Segelquist (*Fish and Wildlife Service*), Paul Shaffer (*ManTech Environmental Technology, Inc.*), Bill Sipple (*EPA Wetlands Division*), Art Spingarn (*EPA Region 3*), Michelle Stevens (*Washington Department of Ecology*), Rich Sumner (*EPA Regional Liaison*), Curtis Tanner (*EPA Region 10*), Ron Tuttle (*Soil Conservation Service*), Virginia Van Sickle-Burkett (*Fish and Wildlife Service*), Fred Weinmann (*EPA Region 10*), Milton Weller (*Texas A&M University*), Joy Zedler (*Pacific Estuarine Research Laboratory, San Diego State University*).

Finally, we recognize those who turned our writings into a finished document. This manuscript could not exist without Kristina Miller's exceptional skills in document and graphics production. We especially appreciate her cheerful tolerance of our times of indecision and needs for immediate changes. Linda Chesnut-Korwin's creativity improved the format of the Executive Summary, special insert, Volunteers and Natural Resource Monitoring, and Tables I, 4-1 and 4-2. Graphic artist, Linda Haygarth produced the figures that needed a "human touch", i.e., could not be made on the computer.

Table I. **Recent publications on wetlands.** Books, proceedings, and reports published since the preparation of **Wetland Creation and Restoration: The Status of the Science** (Kusler and Kentula 1990a) are listed and described below. This list is presented as a guide to the recent literature and is not intended to be comprehensive. Numerous other references are discussed throughout this document and are listed in the References.

Coastal Marshes: Ecology and Wildlife Management
Chabreck (1988)
This book describes the coastal marshes of the United States, their form, functions, ecological relationships, wildlife value, and their management for wildlife. The marshes of the northern coast of the Gulf of Mexico are emphasized.

The Ecology and Management of Wetlands, Volumes 1 & 2
Hook et al. (1988)
This book contains the proceedings of a symposium held at the College of Charleston in 1986. The contributions have been organized to focus on (1) the resource and the basic biology and ecology of wetland plants, animals, soils, hydrology and their values and interactions, and (2) the practicality of applying such information to protect and manage the wetland resource.

Wetland Modelling
Mitsch et al. (1988)
This volume is a statement of the state of the art of modelling approaches for the quantitative study of wetlands. Chapters present different aspects of wetland modelling or a case study characteristic of wetland modelling. Modelling approaches for a wide variety of wetland types are included, as well as models with an emphasis on wetland hydrology, biological productivity and processes and wetland management, and for designing and summarizing large scale research projects.

Proceedings of an International Symposium: Wetlands and River Corridor Management
Kusler and Daly (1989)
The papers in this volume address river and stream corridor management, including the adjacent riverine and estuarine wetlands, from a natural systems protection and restoration perspective. Most of the papers were presented at the International Symposium: Wetlands and River Corridor Management which was held in Charleston, South Carolina, in July 1989.

Wetlands Ecology and Conservation: Emphasis in Pennsylvania
Majumdar et al. (1989)
Wetland experts in the field address a variety of topics on geologic, chemical and biological aspects of wetland ecology. Several chapters are devoted to wetland preservation, and also to increasing our wetland resources through enhancement and mitigation. Important wetland issues such as endangered species, mitigation, and pollution abatement are discussed in detail. The book explains the complexities of protecting wetlands, from delineating boundaries to applying for a permit, to restoring a degraded wetland.

Freshwater Wetlands and Wildlife
Sharitz and Gibbons (1989)
This volume is a product of the Freshwater Wetlands and Wildlfe Symposium held in Charleston, South Carolina, in March 1986. It addresses issues related to natural, man-managed, and degraded ecosystems. The first section deals with the functions and values of wetlands, including their use as habitat, role in trophic dynamics, and basic processes. The second section discusses their status and management, including techniques for assessing value, laws for protection, and plans for management.

An Approach to Improving Decision Making in Wetland Restoration and Creation

Table I. (continued)

Northern Prairie Wetlands
van der Valk (1989)

This book is primarily a review of the ecology of palustrine and lacustrine wetlands in the northern prairie region, i.e, the prairie pothole region of the United States and Canada plus the Nebraska sandhills. It developed out of a symposium held in 1985 at the Northern Prairie Wildlife Research Center of the U.S. Fish and Wildlife Service in Jamestown, North Dakota.

Buffer Zones for Water, Wetlands, and Wildlfe in East Central Florida
Brown et al. (1990)

This report presents a method for estimating buffer sizes necessary in counties in east central Florida to achieve wetland protection through minimization of groundwater drawdown in wetlands, minimization of sediment transport into wetlands, and protection of wildlife habitat. Standards and criteria, minimum buffer requirements, and site-specific measurements that could be used to determine buffers on a site-by-site basis are proposed.

Ecological Processes and Cumulative Impacts: Illustrated by Bottomland Hardwood Wetland Ecosystems
Gosselink et al. (1990)

This book presents the results of three workshops convened by the EPA and facilitated by a team from the National Ecology Research Center of the U.S. Fish and Wildlife Service to solicit expert advice on bottomland hardwood forest ecosystems. The reports from the workshops are summarized in chapters on hydrology, soils, water quality, vegetation, fisheries, wildlife, ecosystems processes and cumulative impacts, and culture/recreation/economics.

Forested Wetlands
Lugo et al. (1990)

This volume of Elsevier's series, Ecosystems of the World, is intended as an introduction to the subject of forested wetlands. The first part reviews available information on the structure and function of forested wetlands and is strongly biased toward forested wetlands in the Caribbean and the United States. The second part presents case studies and descriptions of forested wetlands from other parts of the world.

A Guide to Wetland Functional Design
Marble (1990)

This guidebook presents design guidance that was developed by working each of the function keys in the Wetland Evaluation Technique (WET) backwards to identify the predictors which generate a "high" rating. The guidebook discusses conceptual design, site selection, and site design.

A Manual for Assessing Restored and Natural Coastal Wetlands, With Examples from Southern California
Pacific Estuarine Research Laboratory (1990)

The manual summarizes reference data from several Southern California wetlands, presents a case study from San Diego Bay, and promotes the standardization of assessment methods for restored and natural wetlands by recommending specific sampling and measurement techniques.

Synthesis of Soil-Plant Correspondence Data From Twelve Wetland Studies Throughout the United States
Segelquist et al. (1990)

This report synthesizes the information collected for the U.S. Fish and Wildlife Service in a series of 12 studies designed to describe the relation between soils and vegetation in wetlands located in 11 states throughout the U.S.

Table I. (continued)

Mitigation Site Type Classification System (MiST)
White et al. (1990)
A system for evaluating sites to be restored as forested wetlands based upon the condition of key site factors controlling productivity. Degree of monitoring required is keyed to the site classification.

Wetlands of North America
Niering (1991)
This beautifully illustrated book provides an introduction to wetlands. It is divided into four chapters representing each of the major wetland types—freshwater marshes, coastal wetlands, swamps and riparian wetlands, and bogs and peatlands.

Estuarine Habitat Assessment Protocol
Simenstad et al. (1991)
The Protocol was developed in response to the need for procedures that quantitatively assess the function of estuarine wetlands and associated nearshore habitats for fish and wildlife. It is based on the use of systematic, on-site measurement of habitat function by assessing the attributes of the habitats identified as being functionally important to fish and wildlife.

Creating Freshwater Wetlands
Hammer (1992)
This book is an attempt to organize and present information on methods to create or restore freshwater wetlands accumulated by wetland scientists and managers during the last 50 years.

Restoration of Aquatic Ecosystems: Science, Technology, and Public Policy
National Research Council (1992)
This volume examines the prospects for repairing the damage society has done to the nation's aquatic resources: lakes, rivers and streams, and wetlands. It outlines a national strategy for aquatic restoration, with practical recommendations covering both the desired scope and scale of projects and needed governmental action. Case studies of aquatic restoration activities throughout the country are presented.

Chapter 13: Wetland Restoration, Enhancement, and Creation
U.S.D.A. Soil Conservation Service (1992)
Wetland restoration, enhancement, and creation is now the focus of Chapter 13 of the Soil Conservation Service's Field Handbook. This chapter replaces and expands upon the material that formerly described the construction of dikes and levees.

An Approach to Improving Decision Making in Wetland Restoration and Creation

CHAPTER 1

Introduction

The management of the Nation's wetland resource is characterized by controversy. There is agreement that wetlands are an important component of the landscape, providing a variety of ecological, social, and aesthetic benefits. There is also agreement that over half of the resource has been lost due to conversion to other uses (Tiner 1984), and that the resource needs protection. There is, however, disagreement as to how wetland protection is to be accomplished.

Efforts to protect wetlands are increasing as are the economic pressures to convert them. Government agencies attempt to balance the needs for protection and development through their management decisions as to when and where wetland alteration will be permitted and compensatory mitigation will be required. At the same time, parties immersed in wetlands issues expect some level of predictability in wetland management and regulatory decisions. To establish consistency in the management process, we need a coherent framework for linking current wetland management and regulation. Creative new initiatives are also necessary to address significant voids in protection, inefficiencies in existing programs, and counterproductive public actions and incentives (The Conservation Foundation 1988).

A comprehensive program of wetland management and regulation requires information on the ecological functions of wetlands individually and in the local landscape, and on our ability to create and restore wetlands. Research projects implemented by the Environmental Protection Agency's (EPA) Wetlands Research Program (WRP) are designed to supply this information

(Zedler and Kentula 1986, Leibowitz et al. 1992). Specifically, Agency personnel surveyed in the planning process agreed that the key question is: Do restored and created wetlands perform the same ecological functions as natural wetlands? Moreover, efforts to evaluate the success of wetland restoration and creation projects have been complicated by a lack of stated project goals and by a lack of agreement on what constitutes success. To begin the process of defining success in an ecologically meaningful way, we have developed an approach to establishing ecological criteria for wetland restoration and creation based on results of our studies and related current research. The WRP Approach to evaluating wetlands and wetland projects can be used to tailor resource management to meet specific local and regional needs.

TERMS USED

The WRP Approach can be used to guide the evaluation of created, restored, enhanced, rehabilitated, constructed, and, fundamentally, any human-manipulated wetlands, as well as natural wetlands. To avoid listing all possible situations that might apply and to avoid the current problems of multiple meanings for some terms, we use **projects** to refer to all wetlands that were created or were "improved" by human activities for a specific purpose (e.g., mitigation). We use **natural** to refer to wetlands that occur naturally in the landscape. However, there are times when we need to be more specific. In those cases we use **restored** to refer to any manipulation of a site that contains or has contained a wetland and **created** to refer to attempts to construct a wetland in an area that never has contained a wetland. **Population**, statistically speaking, refers to those wetlands of a similar type and size, either natural or projects, occurring within a geographically defined area.

As mentioned above, the word **success** has a number of meanings as it relates to wetland projects. Success of projects is often rated either on the basis of compliance with permit requirements, or on the basis of whether or not the projects were implemented (Quammen 1986). We believe that **success** must be defined in terms of the project objectives, i.e., what is acceptable for a particular project in a specific locale. For some, success is meeting the terms of the contract; for others, replacement of all aspects of a natural system; for others, replacement of some functions to some level. We leave the definition of project objectives, and the associated success, to those planning or regulating the project. Instead, we offer a way to quantify ecological performance and, ultimately, verify that project objectives have been met, however they are defined.

THE WRP APPROACH AND ITS APPLICATIONS

The WRP Approach demonstrates how information from a monitoring program that includes both natural wetlands and those restored and created can be used to develop performance criteria and suggest improvements to the design of future wetland projects. The same information can be used to evaluate the effectiveness of the management strategy being used. Ultimately, information collected over time can be used to evaluate and refine performance criteria and design guidelines, and to ascertain when a project is developing as expected and when corrections are necessary.

Although wetland restoration and creation are central to this document, it is not a step-by-step approach to building wetlands. Instead, our Approach is a framework for the development of ecologically defensible management strategies for restoration and creation that are tailored to local and regional needs. Such a framework would assist managers in identifying which projects proposals and permit applications should 1) be most critically reviewed because of low probability of success; 2) require the most comprehensive checks on project design and implementation because of uncertain probability of success; and 3) require minimal checks on design and implementation of the mitigation project because of high probability of success. The overall framework is flexible enough to be applied in any area and to any wetland type. However, the monitoring techniques and examples presented will transfer most readily to freshwater nontidal wetlands because they have been the WRP's focus.

KEY CONCEPTS

The steps in the WRP Approach are diagrammed in Figure 1-1. Each of the following chapters discusses one or more of the steps and illustrates the concept using data from the WRP's field studies of natural wetlands and mitigation projects. Briefly, the Approach prescribes compiling and analyzing information from the project files. The results of the analysis are then used to select sites for inclusion in a monitoring program. Data collected are analyzed to determine the performance of the projects, and generate performance criteria and design guidelines for future projects.

Populations

The overall strategy centers around comparisons of samples of populations of natural wetlands and projects. We feel that it is important to consider the variability of ecosystems when making management decisions, especially when setting criteria for performance. Case studies of single sites, and comparisons of pairs of sites do not provide information that can be extrapolated with known certainty to the population as a whole. Therefore, we use the term population in a statistical sense.

The samples of both populations being compared are used as reference sites, i.e., the natural wetlands and wetland projects being sampled are each a group of reference sites. The natural wetlands are a reference against which the development of the projects is judged. The older projects are a reference

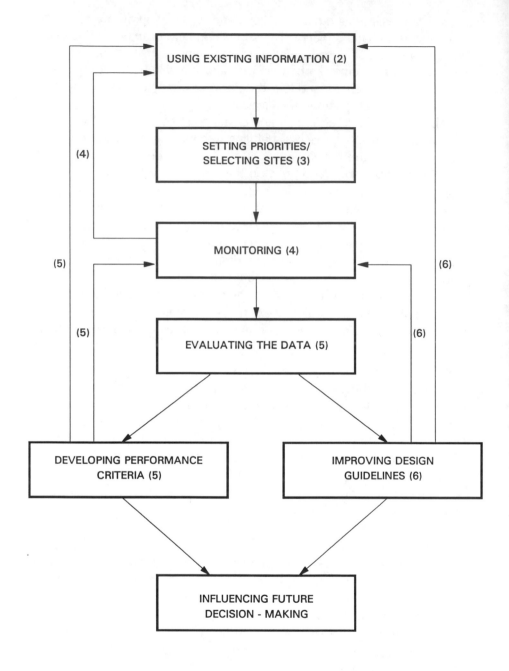

Figure 1-1. The steps in the WRP Approach for using quantitative information to support decision making; in particular, for developing performance criteria and evaluating design guidelines for restored and created wetlands. The numbers in parentheses indicate the chapter in which the concept diagrammed is discussed.

An Approach to Improving Decision Making in Wetland Restoration and Creation

against which the development of similar, newer projects is judged. Both are important. The natural wetlands are used to establish how well goals are being met; the older projects are used to verify that other projects are developing as expected or to detect the results of changes in design.

Setting

The ecological setting of the wetlands is considered in defining the populations and stratifying the samples (see Chapter 3 for details). Brooks and Hughes (1988) suggested Omernik's (1987) ecoregions as a framework for the selection of reference wetlands because they reflect regional patterns of land use, land surface form, potential natural vegetation, and soils. Moreover, they were shown to be an appropriate framework for the selection of reference sites in studies of streams in Arkansas (Rohm et al. 1987), Ohio (Larsen et al. 1986, Whittier et al. 1987), Oregon (Hughes et al. 1987, Whittier et al. 1988), Colorado (Gallant et al. 1989) and Wisconsin (Lyons 1989). Because wetlands and their functions are affected by many of the same factors important to the function and quality of streams, we adopted Omernik's (1987) ecoregions as the regional framework for the selection of sites. In addition, we account for potential effects of land use and position in a watershed in site selection by grouping wetlands in similar land use settings and watershed positions. Comparison of projects with natural wetlands occupying similar landscapes and, therefore, having potentially similar ecological conditions, ensures that what is expected of a wetland project is within the bounds of possible performance given the setting. This framework follows a rationale previously outlined by Bedford and Preston (1988).

Performance Curves

A key analytical tool of the Approach is the performance curve. The performance curve documents the development of the ecological function of projects over time relative to levels of function of similar natural wetlands. Figure 1-2 illustrates one form that a curve could take in an idealized, hypothetical example. Fundamentally, the ability to replace wetland function and the way in which that replacement occurs depend on the type of wetland, the function to be restored or created, and, in the case of restoration, the type of impact that altered the original wetland (Kusler and Kentula 1990b).

Key aspects of the performance curve are labelled on Figure 1-2 which illustrates one possible scenario, in this case a restoration. A is the mean level of wetland function of the restored sites prior to the implementation of the project. A>0 in the case of restoration where there is some level of wetland function prior to the project and A=0 in the case of a creation where there is no level of wetland function prior to the project. B is the mean level of function after the restored wetlands have fully matured. The difference between A and

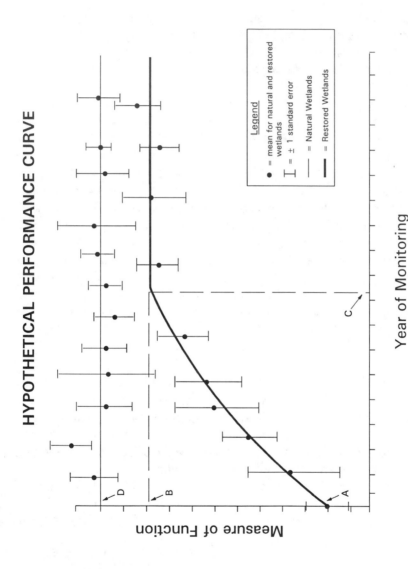

Figure 1-2. Hypothetical performance curve illustrating the comparison of natural wetlands and projects (in this case restored wetlands) of the same type and similar size in the same land use setting relative to a measure of wetland function. A is the mean level of function prior to restoration. B is the mean level of function of the projects when mature. C is the time needed for the projects to mature, i.e, for the level of function to stabilize. D is the mean level of function of the natural wetlands over the time monitored.

B, therefore, represents the amount of function gained due to the construction of the projects. C is the time needed for projects to mature and reach a stable level of function. D is the mean level of function of the natural wetlands over the time monitored. D minus B is the difference in the mean level of function of the natural wetlands and that of mature projects. The values of A, B, C, D and D minus B, as well as the shape of the curve, provide information relevant to wetland management. The shape of the curve can be used to decide when to monitor. For example, inflections in the curve would indicate a change in the rate of development, so monitoring immediately before and after you expect the change could be used to see if the change occurred as expected. The curve can also be used to decide when a project has met or should meet its goals. For example, this could be when the project is C years old, OR has a level of function equal to B plus or minus one standard error of the mean, OR passes a critical stage in development, i.e., experience has demonstrated that projects reaching a certain point on the curve mature as expected.

Using a similar approach, research carried out by EPA's Office of Policy, Planning, and Evaluation in cooperation with the University of Maryland's Center for Environmental and Estuarine Studies focused on an integrated framework for evaluating the cost and performance of wetland creation and restoration projects (King 1991a and King 1991b). King (1991a) demonstrates how the shape of the performance curve for a given site can be affected by the characteristics of the creation or restoration project which are often determined by the amount of resources committed to the project. In follow-up work, King (1991b) argues that financial incentives in wetland mitigation markets reward low cost, not high quality wetland restoration, and account for the relatively poor performance of many restoration projects.

Indicators

Indicators, generally speaking, are variables so closely associated with particular wetland functions that their presence or value is symptomatic of the existence or level of function. Generation of the performance curves depends on having reliable indicators of wetland function. If the indicators are going to be used with any regularity, they need to be sensitive enough to determine functional performance in a reasonable amount of time at a reasonable cost both in dollars and in damage to the wetland. Measures of wetland structure, e.g., site morphology or species present, are readily available and more often meet the requirements of expediency and economy than do direct measures of function. Therefore, measures of structure are frequently used as indicators of wetland function. Some of the typical measures of wetland structure become, when measured over time, measures of function. An example of this is diversity. When measured once, diversity describes the community of organisms of interest at one point in time and is a measure of structure. When measured over time, it documents the system's ability to maintain a level of diversity, which is a function.

Indicator development in wetland science has focused primarily on variables that signify a wetland is present, not specific wetland functions. Although verifying that wetland function exists is the ultimate goal, it is important to establish that a project is, indeed, a wetland and that it maintains the characteristics of a wetland over time. Therefore, we recommend that at least one variable measuring each of the three parameters (wetland hydrology, hydrophytes, and hydric soils) that indicate the presence of a wetland be included in any monitoring program. At the minimum, you will be establishing that the wetland functions associated with a particular type of wetland may exist at some level, since the characteristics of that wetland are present.

We envision that a set of performance curves will be produced over time for each indicator or function measured. What is measured is determined by the goals of the resource management program and the specific projects (Chapter 4). The curves are then examined to identify patterns that can be used as performance criteria, to track a project's development (Chapter 5), and to improve project design (Chapter 6).

Summary

The WRP Approach is a framework for collecting and using information on populations of restored and created wetlands in a given locale. By building a quantitative database of information on what works, what constitutes a successful wetland project, what does not work, what causes damage or loss, the Approach provides the scientific information necessary to help resource managers make decisions that will work and that are defensible.

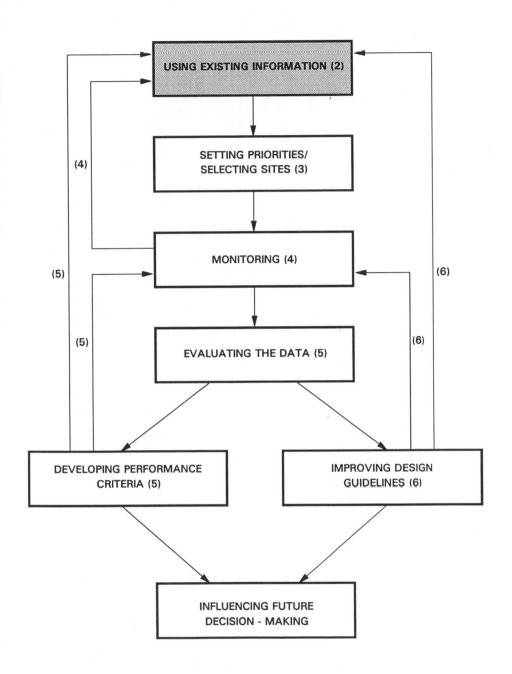

CHAPTER 2

Using Existing Information

A deluge of wetland project files exists in many federal, state, and private agencies involved with wetland regulation and management. The details of the final project agreements, however, are seldom documented or accessible. Decisions typically are made on a case-by-case basis without benefit of quantitative information on how previously granted projects relate to the current proposal or how they affect the status of wetlands in the region (Kentula et al. 1992, Holland and Kentula In press, Sifneos et al. In press(a), Sifneos et al. In press(b)). Data in the project files, therefore, must be updated, compiled, analyzed, and reported if the information is to be reflected in management decisions.

The information in the project files, once in an accessible format, can be used to determine wetland types and locales at risk, to evaluate wetland management practices, and ultimately to influence policy. For example, analysis of previous trends in permitting can reveal locales, wetland types, and functions subject to the most intense permitting activity. With knowledge of such trends, permitting agencies can take actions to avoid additional losses in wetland numbers, types, functions, and area.

Although wetlands are constantly being lost to natural forces and human activities, such as erosion, drainage, and land-clearing, regulation through permitting is one mechanism by which agencies can influence the wetland inventory. Periodic assessments of the cumulative impacts of various permitting systems (e.g., Clean Water Act Sections 404 and 401, Rivers and Harbors Act Section 10, and state regulations) on wetlands are essential for determining the overall effects of permitting on the wetland resource. Unfortunately, the quality of the documentation of management decisions has been inadequate for reliable descriptions of trends in the status of the resource or for evaluation of management strategies. For example, we analyzed databases containing information from portions of the Section 404 permit record from the 1970s and 1980s for eight states (Oregon, Washington, Louisiana, Mississippi, Alabama, Texas, Arkansas, and California) (Table 2-1). In all eight states, information on the impacted wetlands and mitigation projects was either lacking or of poor quality. Approximately 40% of the impacted wetlands and mitigation projects in California lacked acreage data; therefore, area trends reported for the state might be misleading (Holland and Kentula In press). Furthermore, information on project completion dates was inadequate for all eight states. In Louisiana, only 3% of the mitigation projects had completion dates listed in the permit records (Sifneos et al. In press(a)). A large percentage of permits issued in several states lacked specific locations for the wetlands. A better assessment of the effects of permitting on wetlands would be possible if record keeping were improved and standardized. In particular, this would allow consideration of the cumulative effects of individual permit decisions on the wetland resource.

MINIMUM INFORMATION NEEDED

Although detailed information on all permits and projects should be kept in the files, it is imperative that a subset of this information be compiled, en-

Table 2-1. Summary of the Section 404 permit databases compiled by EPA's Wetlands Research Program. IMP=number of wetlands impacted; COMP=number of compensatory wetlands.

State	Information Compiled	# Permits	# Wetlands	
			IMP	COMP
OR	All permits requiring mitigation 1977-January 1987	58	82	80
WA	All permits requiring mitigation, 1980-1986	35	72	52
TX	All permits involving freshwater wetlands and requiring mitigation, 1982-1986	46	71	72
AR	All permits involving freshwater wetlands and requiring mitigation, 1982-1986	7	8	9
AL	All permits involving freshwater wetlands, 1982-August 1987	18	28	23
MS	All permits involving freshwater wetlands, 1982-August 1987	10	11	6
LA	All permits involving freshwater wetlands, 1982-August 1987	226	258	116
CA	All permits requiring mitigation, 1971-November 1987	324	368	387

An Approach to Improving Decision Making in Wetland Restoration and Creation

tered into a database, and periodically analyzed and reported to identify trends in decision making and areas at risk. The subset should include information such as the specific location of the impacted wetlands and mitigation projects, dates that permits were issued and mitigation projects were begun and completed, wetland types (e.g., according to Cowardin et al. 1979) and areas, functions of the impacted wetlands, objectives of the projects, and summaries of monitoring information. Table 2-2 lists the minimum information that we recommend be compiled for adequate descriptions of trends in permitting activity. Similar information also can be collected for projects implemented outside the permitting process. Ideally, the information compiled and reported would be standardized nationally to facilitate comparisons between states and regions. We recommend statewide standardization as a minimum goal.

The most accurate and comprehensive trends can be identified by compiling information from the historic record, entering it into a database, and analyzing and reporting the results. For example, you could enter all Section 404 permits that required compensatory mitigation in a state into a database to track the effects on the status of the wetland resource in the state. Examination of the historic record can be used to: 1) identify locations with the most intense project activity; 2) identify the wetland types most frequently impacted and used as compensatory mitigation; 3) ascertain trends over time; and 4) select areas for further study.

Compiling the historic record, however, can be extremely time-consuming and costly. As an alternative, we recommend that you start with the present and continue compiling information into the future. Although this approach will not provide you with information on historic trends, you will have a method of quantifying project data and detecting trends in the future. You can always compile the historic information, as resources allow, beginning with the present and working backward a year at a time.

It is essential that the minimum information recommended in Table 2-2 be recorded for all projects and be available to resource managers. Trends in decision making involving wetland projects and their effects on the resource cannot be evaluated unless information is compiled, entered into a computerized database, the data analyzed, and cumulative impacts of individual projects on the wetlands resource assessed. See the following section, "Features of EPA's Permit Tracking System", for insights on data quality assurance, data analysis, and data retrieval.

FEATURES OF EPA'S PERMIT TRACKING SYSTEM

An example of a data management system developed to simplify the process of entering and analyzing the information from permit records is the Permit Tracking System (PTS) (Holland and Kentula 1991). The PTS is a user-friendly, PC based program, designed to track information from three types of

Table 2-2. Minimum categories of data on impacted and compensatory (created, enhanced, preserved, or restored) wetlands recommended for inclusion in a database and data categories found in EPA's Permit Tracking System (PTS) (Holland and Kentula 1991).

MINIMUM	PTS
IMPACTED AND COMPENSATORY WETLAND	

Location   State and county
    Specific location
    Waterbody/river basin
    Land use
	Location State and county
	    Specific location
	    Waterbody/river basin
	    Land use
	    USGS map name and scale
	    Latitude/Longitude
	    Township/Range/Section

Dates Permit Issued
    Construction began/completed
	Dates Permit Issued
	    Construction began/completed

Cowardin wetland types	Cowardin wetland types
Area of the wetlands	Area of the wetlands
Contact	Contact
	Documents available
	Reports
IMPACTED WETLAND ONLY	
Project type	Project type

Functions documented
    Endangered species names
	Functions documented
	    Endangered species names

COMPENSATORY WETLAND ONLY	

	Was mitigation bank used?
	    Name of mitigation bank
	    Money or land?

Compensation type	Compensation type
On-site or off-site?	On-site or off site?
	Were corrections made?

Objectives stated
    Endangered species names
	Objectives stated
	    Endangered species names

Monitoring information
    Do "as-built" plans exist?
    Regular or irregular checks made?
    Items monitored
	Monitoring information
	    Do "as-built" plans exist?
	    Regular or irregular checks made?
	    Items monitored

	Methods of construction

wetland permit systems: Section 404, Section 401, and state. The program includes an option to track data from other permit systems or wetland projects.

We present the PTS as an example of a system that could be used to compile and analyze information from project files (Table 2-2). It is composed of two main components, data entry and query. The PTS simplifies the process of data entry, because in most cases, the user is merely required to check off items, as opposed to entering verbiage. Standardized categories, with definitions, are given for items such as wetland type, project type, and wetland function. Selecting items and entering minimal verbiage eliminates most of the errors typically associated with data entry. The PTS also sorts and prints all the items listed in each category, making it easy to recognize information that has been entered incorrectly. For example, if a list of county names included CENTER and CENTRE, it would be simple to recognize the error in data entry. After data have been entered, corrections, deletions, and additions can be incorporated into the database.

The menu-driven query component of the PTS allows the user to generate queries using the contents of the database (Figure 2-1). The program identifies all possible combinations of queries and compiles the answers, which can be viewed on the screen, copied to disk for conversion to tables and figures, or printed as hard copy.

The PTS not only eliminates the potential errors inherent to querying in other software packages, but also substantially reduces the time required for analyses. For example, analysis of the Oregon database using dBase III+ took approximately three weeks. When we tested the PTS by reanalyzing the Oregon data, analysis time was reduced to only three days. Furthermore, analyses using the PTS involve minimal user time. For example, using the PTS to calculate the number of impacted wetlands and mitigation projects for each wetland type entails setting up only one query, which takes approximately two minutes. The computer can then be left unattended as the PTS calculates the results. Traditional software packages require the user to enter a query for each wetland type, entailing substantially more user time.

Although it was designed to track information from the permit record, the PTS, or a similar data management system, could be used to track projects implemented outside the permitting process. The key concept is that information from project files must be compiled and entered into a database, so the data can be analyzed and made accessible to those making resource management decisions.

INCORPORATING ADDITIONAL INFORMATION

The process of compiling information does not stop when the details of the project plan or permit are documented. All the project information that is collected should be incorporated into the file so that the record is as complete as

1. QUERY: Was there a net change in wetland area as a
           result of Section 404 permitting?

2. QUERY SCREEN FROM PTS:

CASE_T/F	CASE		AREA_L	AREA_H
.T.			0.0	9999.9
.F.	Impacted			
.F.	Created			
.F.	Enhanced			
.F.	Preserved			
.F.	Restored			

A "T" (True) in the top row
indicates that information is
needed for all subsequent rows.

The range columns define
the upper and lower limits
for area.

3. RESULTS SCREEN FROM PTS:

CASE		AREA (acres)
Impacted		90.0
Created		12.0
Enhanced		4.0
Preserved		10.0
Restored		24.0

4. RESULTS:

Area compensated:	12.0 + 4.0 + 10.0 + 24.0 =	50.0 acres
Area impacted:		-90.0 acres
Net change in area		-40.0 acres

Figure 2-1. Examples of the query and results screens from the Permit Tracking System (PTS)
(Holland and Kentula 1991) generated to answer the question: "Was there a net
change in wetland area as a result of management decisions?"

An Approach to Improving Decision Making in Wetland Restoration and Creation

possible. A seemingly unimportant fact can become a key piece of data. For example, it is important to include monitoring data in the project file. You can then use it to determine if the project is in compliance with the permit specifications and project objectives. The monitoring information from all similar projects can be entered into a database and analyzed to determine if the projects are functioning as planned and if management objectives are being met (see Chapters 4, 5, and 6). Information that is related to the overall management strategy should also be entered into a database, analyzed, and the results used to make decisions and evaluate the strategy. If all pertinent project information is available for use in decision making, management strategies and regulatory decisions will be based on the most up-to-date, scientifically defensible information.

REPORTING THE INFORMATION

Regular reports summarizing information in the project files are necessary to provide a comprehensive assessment of the wetland resource for areas of interest, such as states, regions, watersheds, or ecoregions. Furthermore, reporting provides a mechanism for assessing risks to wetlands. For example, trends, such as the loss of certain wetland types, can be identified with regular reporting. Once trends are identified, actions can be taken to avoid losses in wetland numbers, types, functions, and area. Dissemination of the reports to local, regional, and national authorities is critical if information in the reports is to be reflected in management decisions. Finally, regular reports will provide a mechanism for policy makers and planners to receive the information in a usable format.

Our analyses of Section 404 permitting during the 1970s and 1980s revealed several notable trends that could be used by resource managers in evaluating the effects of permit decisions. In most of the states studied, more wetland area was destroyed than was required to be created or restored, resulting in net losses in wetland area. Furthermore, less than 55% of the permits for all eight states analyzed required that the mitigation projects be monitored by at least one site visit; the range was from 0 monitored in Arkansas to 52% monitored in Texas (Sifneos et al. In press(b)) (Figure 2-2). The wetland types of the mitigation projects often differed from the wetlands destroyed, resulting in net losses in area for certain wetland types. For example, as a result of permits requiring compensatory mitigation, palustrine forested wetland was the wetland type subject to the greatest loss in area in California (-143.9 ha) (Holland and Kentula In press) and Louisiana (-414.3 ha) (Sifneos et al. In press(a)), whereas palustrine emergent wetland was the type subject to the greatest loss in area in Oregon (-15.0 ha) and Washington (-9.7 ha) (Kentula et al. 1992) (Figure 2-3). Permitting activity involving compensatory mitigation was concentrated near urban areas in several states. In Oregon, it occurred near Portland and Coos

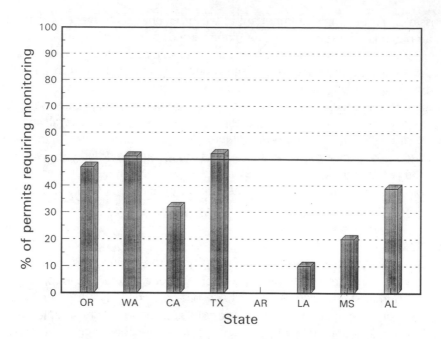

Figure 2-2. Comparison by state of the percent of the Section 404 permits requiring compensatory mitigation that specified monitoring the project with at least one site visit.

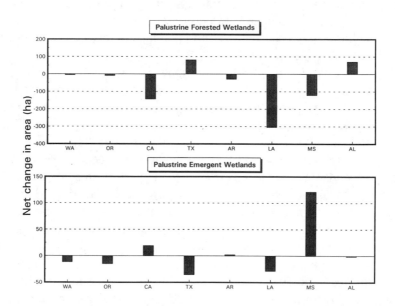

Figure 2-3. Comparison by state of the net change in area of palustrine forested wetlands and palustrine emergent wetlands involved in Section 404 permits requiring compensatory mitigation over the time period analyzed (see Table 2-1). The data were obtained from the Section 404 permit record.

An Approach to Improving Decision Making in Wetland Restoration and Creation

18

Bay, and in Texas it was clustered around the Dallas-Fort Worth metropolitan area. In addition, Section 404 permitting destroyed endangered species habitat in most states evaluated. The trends in permitting described above were obtained by compiling and analyzing portions of the Section 404 permit records. However, unless such trends are reported, the effects of management decisions on the wetland resource will remain unknown.

SUMMARY

The information in wetland project files must be updated, compiled, analyzed, and reported, if the data are to be of use in protecting the wetland resource. Trends and patterns in the information can be used to identify important issues for further examination, which in turn, can guide management decisions. For reliable descriptions of the effects of management decisions, complete and accurate information is required for all projects. Only with improved documentation and regular reporting can we credibly assess the cumulative effects of decisions involving individual or small groups of wetlands on the resource.

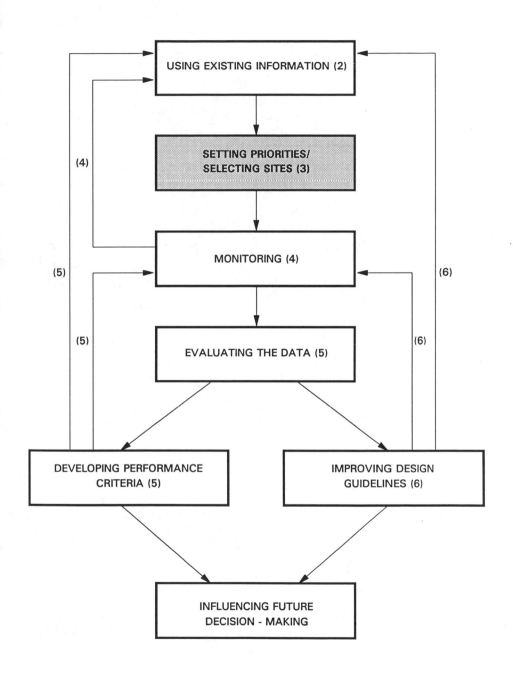

CHAPTER 3

Setting Priorities and Selecting Sites

The ability to implement the WRP Approach successfully hinges on good planning. Often the resources of an agency or organization are limited, so we recommend that priorities be set before instituting a monitoring program. In this chapter we discuss how to target sampling to critical areas and wetland types, and present a procedure for selecting both natural wetlands and projects to monitor. We use our experiences studying mitigation projects to illustrate the process of setting priorities and selecting sites. However, note that the application of the process described is not limited to mitigation projects, and that we present but one way that a statistical population of wetlands can be defined.

To simplify the presentation, this section is written as though the Approach will be implemented in one area and with one group of projects (e.g., one wetland type and size class). In fact, the procedures can be used to identify more than one area or group of projects to be monitored. The scope of the application of the Approach is determined by the needs of the agency or organization, and, ultimately, the status of the wetland resource.

DECIDING ON A SAMPLING STRATEGY

The hypothetical performance curve described in Chapter 1 displays the changes in function over time in wetland projects as compared to similar natural wetlands. Wetland function is typically measured using an indicator. The curve can be generated in two ways depending on how you sample. One method is to follow the development of similar aged projects by repeatedly sampling the same projects and natural wetlands over time (Figure 3-1a). The other is to gather data from projects and natural wetlands at one time, documenting project development by sampling projects representing a range of ages (Figure 3-1b). The latter is how we have generated performance curves to date.

The sampling approach chosen will depend on the history of wetland restoration and creation in your area. If you are just beginning to construct wetland projects, or if a number of projects are constructed every year or two, you may want to use the first approach and follow groups through time. If wetland restoration and creation has been going on for some time, you may want to use the second approach and sample projects that represent a range of ages. This approach has the advantage of potentially generating a large portion of the performance curve at once. However, the design of the oldest projects may be quite different from that of the newest projects. The effects of the different designs may confound the results so that the pattern of project development is not apparent.

Identifying Priority Areas

We recommend that monitoring efforts be targeted to areas at greatest risk. Areas at risk are those where the greatest wetland losses in terms of area, ecological function, and/or value have occurred, are occurring, or are anticipated to occur. In this case, value represents the benefits of the wetland that are realized or recognized by society and includes uniqueness and rarity (Leibowitz et al. 1992). In addition, the areas should have a high probability of producing useful information. However, in some cases an area with a low probability of producing useful information will be favored because the area or the wetland type is so important that any information obtained will be of great benefit. Depending on the causes of the losses, the areas chosen will probably be places where there is an abundance of Section 404 and associated permitting activity. Consequently, these areas will be places where many wetland restoration and creation projects are occurring or will be occurring.

One way to identify the areas at risk is to survey the personnel involved in wetland management and, in particular, permitting. Unfortunately, because of the high turnover in regulatory personnel associated with permitting, this approach will not always provide the best answers. In the short term, however, a decision based on the "institutional memory" will probably identify current problems and allow implementation of a monitoring program.

Examination of actual records of wetland losses, restoration efforts, mitigation projects, and growth and development is probably the best way to select areas at risk. Chapter 2 describes a system that can be used to compile information for this purpose. However, if the information is not available in an easily retrievable form, it will probably take a major effort to compile materials

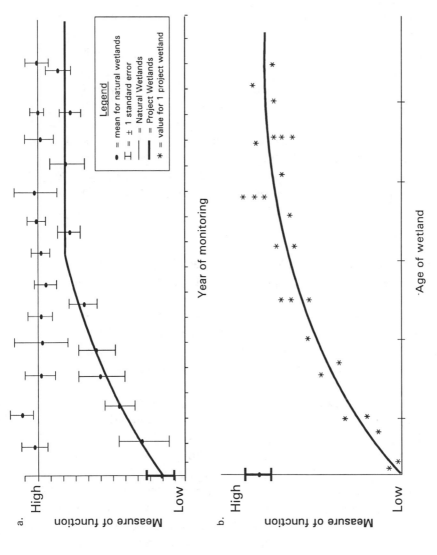

Figure 3-1. Hypothetical performance curves. a) Example of hypothetical performance curve where the same samples of populations of projects and natural wetlands are followed Through time. b) Example of hypothetical performance curve where measurements are taken at one time from a sample of a population of projects of various ages and a sample of a population of natural wetlands.

from files into a computerized database. Therefore, the time and resources needed to collect and organize the information must be taken into account in planning. For instance, it may be more expedient to concentrate on a portion of the record recommended by local staff rather than delay implementing the monitoring program while the entire record is compiled and analyzed. The FWS's National Wetlands Inventory (NWI) reports (e.g., Frayer et al. 1989) are also excellent sources of information on trends in wetland area for the locales for which they have been produced.

The permit database compiled for Oregon (Abbruzzese et al. 1988, Kentula et al. 1992) and similar information found in Florida (Brown 1991) were used to identify mitigation projects for two of our field studies. For example, the record of permits issued in Oregon from January 1977 through January 1987 indicated that 31% of the permits requiring compensatory mitigation involved wetlands in the Portland Metropolitan Area (Kentula et al. 1992) (Figure 3-2). Because Portland continues to be a major growth area, pressures to develop wetlands will probably escalate. Therefore, the Portland Metropolitan Area is considered an area at risk and a priority for management. This clustering of permit activity is not unusual. Figure 3-3 illustrates the patterns we identified in California (Holland and Kentula In press) and Louisiana (Sifneos et al. In press(a)).

SELECTING SITES

Once the area at risk is identified, the next steps are to define the appropriate populations of wetlands to sample and to select a representative sample of sites from each. In the following sections we will describe how to: 1) define the population of projects to be sampled; 2) use information on project location to define the boundaries of a study area; 3) define the population of natural wetlands to be sampled in terms of the characteristics of the population of projects; and 4) finalize the list of projects and natural wetlands to be sampled.

Defining the Population of Wetland Projects to Sample

How you define the population of projects to be monitored will influence the definition of the population of natural wetlands to be sampled as well as the choice of measurements, the timing of sampling, and virtually every aspect of the monitoring scheme. As a rule, document the discussions and decisions that occur during planning. In particular, you should record what was considered in defining the populations of projects and natural wetlands and the outcome of these decisions. The data used in making the decisions should also be included in the documentation. Sometimes the decision steps that lead to an approach or justify a choice of measures are forgotten over the course of a study. Often such information is key to guiding the analysis of data and interpretation of the results.

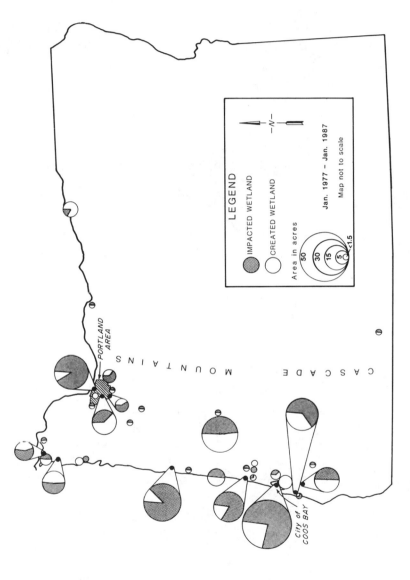

Figure 3-2. Locations of wetland impacts and creations in Oregon that occurred between January 1977 and January 1987 as a result of Section 404 permits requiring compensatory mitigation. Area of wetland impacted and created is expressed in acres. The size of the circle represents the sum of the area impacted and created at each location. Note that 20% of the total area impacted and 18% of the total area created were located in the Portland Metropolitan Area.

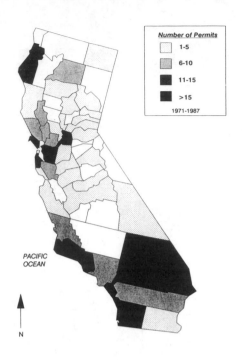

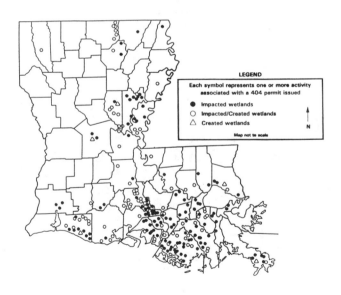

Figure 3-3. Patterns of Section 404 permitting in California and Louisiana. a) The number of permits requiring compensatory mitigation from 1971-1987 is illustrated by county for California. Other patterns in permitting in California are discussed in Holland and Kentula (In press). b) Locations of permitted activity involving freshwater wetlands from January 1982 - August 1987 are illustrated for Louisiana. Other patterns in permitting in Louisiana are discussed in Sifneos et al. (In press(a)).

An Approach to Improving Decision Making in Wetland Restoration and Creation

Start by obtaining a list of all the projects located in the area at risk. For example, you could obtain a list of all the mitigation projects in the area, or access the files from which such a list can be generated from state permitting agencies or the U.S. Army Corps of Engineers' (COE) District Offices. Also, the Soil Conservation Service should be able to provide information on restorations done under the 1990 Farm Bill (Food, Agriculture Conservation and Trade Act of 1990 (P.L. 101-624)); the FWS, on restorations done under the Waterfowl Reserve Program.

Before spending any time "digging" through files, take some time to define the types of information that would be helpful in organizing sites into meaningful groups. This increases the probability that you will find the information you need to make a decision as to whether a site is part of the population of interest. Also, knowing what you want will reduce the number of trips to the files. Figure 3-4 is a form we have used for compiling information we used in site selection. At minimum you will need to know the location of the project, the wetland type and size, and the property owner or a contact for the project.

Because much of our Approach focuses on the development of projects over time, it is advantageous to know when construction was completed to determine the age of the project. However, we have found that this information is often hard to obtain. Knowing the actual date a project was completed would be ideal, but the general time of year is also helpful as is the year of construction (e.g., fall 1987).

Examine the list of projects to identify populations of projects. For example, Table 3-1 lists the freshwater mitigation projects in Portland, Oregon, required in permits issued by the COE and the Oregon Division of State Lands from January 1987 through January 1991. From this information we decided to concentrate on the most frequently occurring type of project—a wetland that was primarily emergent marsh and open water (Figure 3-5). This gave us a reasonably large pool of sites that represented a range of sizes and ages and defined the first characteristic of the population to be sampled—wetland type.

Depending on the size of the population, either the entire group or a random sample of the sites can be used. As you collect data, the variability in the measurements taken will indicate whether the number of wetlands sampled is adequate. If the variability is large, the sample size may need to be increased to improve the precision of your estimates. On the other hand, if the sites are homogeneous, you may be able to decrease the number of sites sampled and save resources. Although it is best to have a large number of projects to choose from, the number of sites available should not be a constraint in implementing the Approach. For example, we detected statistically significant differences between created and natural wetlands for certain variables with a sample of nine created and nine natural wetlands in the Florida Study, and with 11 created and 12 natural wetlands in the Oregon Study.

COE permit number_____

State permit number_____

Date permit issued ___/___/___

Permit Tracking System
COMPENSATORY WETLAND DATA FORM

Form designed by C.C. Holland and R.G. Gibson
ManTech Environmental Technology, Inc.
U.S. Environmental Protection Agency,
Environmental Research Laboratory
200 SW 35th Street
Corvallis, OR 97333

Mitigation type--Select [1]
O Created   O Enhanced   O Preserved   O Restored

State ____ County_____   Acres____.__
State ____ County_____   Acres____.__

TOTAL_____.__

Township & Range_____   Section(s)_____

Latitude/Longitude_____

USGS/NWI map name_____   Scale 1: _____

Select [1]	Water/river body name _____
O Water Body	
O River Body	Specific location _____

Was the mitigation project Off-site or On-site?

Land use--Select [1]	Documents available--Select [0-4]
O Agricultural	O Maps
O Commercial	O Blueprints
O Industrial	O Ground photos
O Natural	O Aerial photos
O Residential	

Date construction began ___/___/___

Date construction completed ___/___/___

Were mid-course corrections made? Yes / No
(Make notes in comments section)

COWARDIN WETLAND TYPE--Select [1-5]		

ACRES		ACRES		ACRES

ESTUARINE

O subtidal aquatic bed ___.__
O subtidal open water ___.__
O subtidal reef ___.__
O subtidal rock bottom ___.__
O subtidal unconsolidated bottom ___.__
O intertidal aquatic bed ___.__
O intertidal beach/bar ___.__
O intertidal emergent ___.__
O intertidal flat ___.__
O intertidal forested ___.__
O intertidal reef ___.__
O intertidal rocky shore ___.__
O intertidal scrub/shrub ___.__
O intertidal streambed ___.__
O intertidal unconsolidated shore ___.__

LACUSTRINE

O limnetic aquatic bed ___.__
O limnetic open water ___.__
O limnetic rock bottom ___.__
O limnetic unconsolidated bottom ___.__
O littoral aquatic bed ___.__
O littoral beach/bar ___.__
O littoral emergent ___.__
O littoral flat ___.__
O littoral open water ___.__
O littoral rock bottom ___.__
O littoral rocky shore ___.__
O littoral unconsolidated bottom ___.__
O littoral unconsolidated shore ___.__

RIVERINE

O tidal aquatic bed ___.__
O tidal beach/bar ___.__
O tidal emergent ___.__
O tidal flat ___.__
O tidal open water ___.__
O tidal rock bottom ___.__
O tidal rocky shore ___.__
O tidal streambed ___.__
O tidal unconsolidated bottom ___.__
O tidal unconsolidated shore ___.__
O lower perennial aquatic bed ___.__
O lower perennial beach/bar ___.__
O lower perennial emergent ___.__
O lower perennial flat ___.__
O lower perennial open water ___.__
O lower perennial rock bottom ___.__
O lower perennial rocky shore ___.__
O lower perennial streambed ___.__
O lower perennial unconsolidated bottom ___.__
O lower perennial unconsolidated shore ___.__
O upper perennial aquatic bed ___.__
O upper perennial beach/bar ___.__
O upper perennial flat ___.__
O upper perennial open water ___.__
O upper perennial rock bottom ___.__
O upper perennial rocky shore ___.__
O upper perennial streambed ___.__
O upper perennial unconsolidated bottom ___.__
O upper perennial unconsolidated shore ___.__
O intermittent aquatic bed ___.__
O intermittent beach/bar ___.__
O intermittent flat ___.__
O intermittent open water ___.__
O intermittent rock bottom ___.__
O intermittent rocky shore ___.__
O intermittent streambed ___.__
O intermittent unconsolidated bottom ___.__

RIVERINE (cont)

O unknown perennial aquatic bed ___.__
O unknown perennial beach/bar ___.__
O unknown perennial flat ___.__
O unknown perennial open water ___.__
O unknown perennial rock bottom ___.__
O unknown perennial rocky shore ___.__
O unknown perennial streambed ___.__
O unknown perennial unconsolidated bottom ___.__
O unknown perennial unconsolidated shore ___.__

PALUSTRINE

O aquatic bed ___.__
O emergent ___.__
O flat ___.__
O forested ___.__
O moss/lichen ___.__
O open water ___.__
O rock bottom ___.__
O scrub/shrub ___.__
O unconsolidated bottom ___.__
O unconsolidated shore ___.__

MARINE

O subtidal aquatic bed ___.__
O subtidal open water ___.__
O subtidal reef ___.__
O subtidal rock bottom ___.__
O subtidal unconsolidated bottom ___.__
O intertidal aquatic bed ___.__
O intertidal beach/bar ___.__
O intertidal flat ___.__
O intertidal reef ___.__
O intertidal rocky shore ___.__
O intertidal unconsolidated shore ___.__

TOTAL AREA ___.__

Figure 3-4. An example of a form that can be used to compile information on wetland projects (Holland and Kentula 1990).

An Approach to Improving Decision Making in Wetland Restoration and Creation

REPORT INFORMATION

Title_____

Author's First Initial_____ Middle Initial_____ Last Name_____

Year_____ Source _____

Content_____

CONTACT INFORMATION

First Initial_____ Middle Initial_____ Last Name_____

Organization_____

Address _____

City_____ State_____ Zip_____ Phone (   ) _____

CONTACT INFORMATION

First Initial_____ Middle Initial_____ Last Name_____

Organization_____

Address _____

City_____ State_____ Zip_____ Phone (   ) _____

CONTACT INFORMATION

First Initial_____ Middle Initial_____ Last Name_____

Organization_____

Address _____

City_____ State_____ Zip_____ Phone (   ) _____

CONTACT INFORMATION

First Initial_____ Middle Initial_____ Last Name_____

Organization_____

Address _____

City_____ State_____ Zip_____ Phone (   ) _____

COMMENTS

Objective:_____
Method:_____
As-built:_____
_____
_____
_____
_____
_____

Figure 3-4. (continued)

Table 3-1. Numbers of freshwater mitigation projects in Portland, Oregon, by wetland type and size required in Section 404 permits issued by the U.S. Army Corps of Engineers and the Oregon Division of State Lands from January 1987 through January 1991.

TYPE	SIZE (Acres)							TOTALS
	0-2	2-4	4-6	6-8	8-10	>10	?	
Marsh	9	5	0	0	0	0	2	16
Pond	24	0	1	0	1	1	12	39
Marsh and Pond	4	1	0	0	0	0	3	8
Marsh and Shrub-scrub	2	2	1	0	0	0	1	6
Marsh and Forested	0	0	0	0	0	0	1	1
Marsh, Shrub-scrub, and Forested	1	0	0	0	0	0	0	1
Marsh, Pond, Shrub-scrub, and Forested	0	1	0	0	1	1	0	3
Marsh, Pond, Aquatic Bed, and Forested	0	1	0	0	0	0	0	1
Marsh, Pond, and Flooded Grassland	0	0	1	0	0	0	0	1
Pond, and Riparian	10	0	0	0	0	0	0	10
Pond, Forested, and Stream Channel	1	0	0	0	0	0	0	1
Riverine Wetland	0	1	0	0	0	0	0	1
Stream Channel	1	0	0	0	0	0	2	3
Creek Bank	1	0	0	0	0	0	0	1
100-year Floodplain	1	0	0	0	0	0	0	1
Mud Flat	1	0	0	0	0	0	0	1
Unknown	3	0	0	0	0	0	10	13
TOTALS	58	11	3	0	2	2	31	107

Figure 3-5. Example of a typical mitigation project sampled in the Oregon Study.

An Approach to Improving Decision Making in Wetland Restoration and Creation

If the number of projects being considered is large, you may want to seek permission for access at this point to prevent wasting time gathering information on sites that are not accessible to the study. See the section on finalizing the lists of sites to be sampled for ideas on how to obtain permission for access.

Defining the Boundaries of a Study Area

In the process of setting boundaries for the study area, the definition of the population of projects is refined to include the concept of ecological setting. The study area can be defined as either the entire area at risk or just a portion of it. The study area should bound a population of projects in as homogeneous an ecological setting as possible. By this we mean that the boundaries should be set to include similar hydrologic, climatic, geologic, or other relevant geographic conditions that influence the nature of the wetlands.

Taking a regional perspective

The boundaries of the study area are set by the following procedure adapted from Abbruzzese et al. (1988). The first step is to examine the distribution of the projects relative to the ecoregion boundaries (e.g., Omernik 1987) to determine which ecoregions to include in the study area. We recommend that sites within the same ecoregion be considered a population. If the area at risk is large, it is advisable first to plot the locations of the projects and the ecoregion boundaries on 1:500,000 scale state maps. Then the overall pattern can be analyzed to identify smaller areas on which to concentrate. If you decide to subdivide the area at risk, consider only those smaller areas in subsequent decisions. The next step is to transfer the locations of the projects and ecoregional boundaries to 1:100,000 scale NWI maps for comparison with U.S. Geological Survey (USGS) topographic or ecoregion maps at the same scale. Examine spatial patterns of relief, hydrographic features, and vegetative cover to identify possible subregions or discontinuities. For example, in the Oregon Study our examination of the topography around the Portland Metropolitan Area showed three distinct subregions (Figure 3-6). The Coast Range extended into the northwest quarter of the city. Small hills dominated the south. Lowlands created by the Columbia and Willamette Rivers occupied the north and west. Most of the mitigation projects were located in the lowlands, therefore the lowlands within the Portland Metropolitan Area defined the boundaries of our study area, and only projects within that area were part of the population studied.

Considering ecological setting

Matching the ecological setting of natural wetlands with that of the projects is a fundamental aspect of the Approach. Our initial tendency was to se-

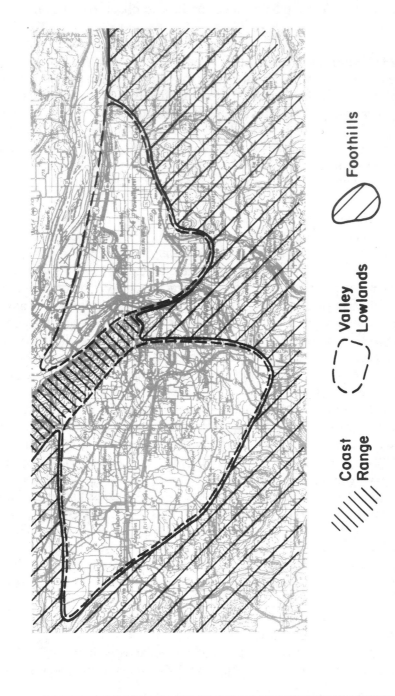

Figure 3-6. An example of how a U.S. Geological Survey topographic map can be used to identify subregions. A 1:250,000 scale map of the area around Portland, Oregon, is coded to indicate the three subregions: the Coast Range, the area dominated by small hills, and the lowlands created by the Columbia and Willamette Rivers.

Coast Range

Valley Lowlands

Foothills

lect the most pristine sites in the ecoregion as the natural wetlands to be used for comparison. For example, Brooks and Hughes (1988) recommended that reference sites should be relatively undisturbed and representative of the region and the population of mitigation sites. However, examination of Section 404 permits requiring wetland creation in Oregon showed the majority of the created wetlands were in or near metropolitan areas (Kentula et al. 1992). Because this pattern also occurred in several of the other states we studied, questions were raised about whether relatively pristine or undisturbed wetlands were legitimate comparisons for projects located in a human influenced setting. We also realize that these questions are related to defining attainable and acceptable performance for the projects. Natural sites chosen from landscapes with limited human influence may not reflect the potential structure and function of projects found in an urban setting. In other words, comparison of wetland projects with natural wetlands located in a similar land use setting and, therefore, exposed to similar ecological conditions, is needed to ensure that what is "expected" of a project is within the bounds of possible performance given the setting.

Land use has become a major part of our definition of ecological setting. For instance, the Oregon Study involved projects and natural wetlands within the Portland Metropolitan Area. While we felt that it was legitimate to compare projects located in an urban setting with natural wetlands in the same setting, we realized that wetlands in urban areas may not be of the same quality as wetlands in other land use settings. We knew it would be valuable to document those differences and use the information to direct wetland protection and restoration to areas with the greatest potential for ecological benefit. In the case of mitigation, such information can be used to avoid compensating for wetland losses with projects of limited ecological value due to their location. In addition, we realized that the information could be used to identify how particular land uses impact wetlands and how to buffer the systems from these impacts. The nature of the interactions among various anthropogenic factors and between anthropogenic and natural variables is a legitimate ecological research topic and one of increasing importance. Knowledge of the relative influences of urban and natural environmental forces on ecosystem function is fundamental to our understanding of ecosystems and the impacts of human activities on them. The necessity of such information was acknowledged in a special feature on urban gradients published in a recent issue of Ecology (Volume 17, Number 4, 1990).

We recommend that the populations of wetland projects and natural wetlands be stratified by land use setting so that the natural wetlands represent the various land uses surrounding the wetland projects. In the Florida Study, we used the Landscape Development Intensity (LDI) index to quantify the development intensity surrounding the wetlands and to stratify the sample (Brown 1991).

Defining and Sampling the Population of Natural Wetlands

In the Oregon Study the population of natural wetlands sampled was defined in terms of the population of projects, (i.e., palustrine emergent marshes with open water that were less than or equal to one hectare in size and located in the Portland Metropolitan Area within the lowlands created by the Columbia and Willamette Rivers (see Figure 3-5). We randomly selected natural wetlands from the population defined using a procedure adapted from Abbruzzese et al. (1988). An overview of the procedure is given below, for additional details see Abbruzzese et al. (1988) and Brown (1991). Certainly this and other methodologies can be adapted to meet your individual needs.

In our procedure, first we overlay a grid, with each cell representing 260 ha, on a 1:100,000 scale USGS topographic map marked with the study boundaries. After sequentially numbering the cells falling within the study area, we transfer the numbered grid to 1:100,000 scale NWI maps. The order in which to sample the numbered cells is determined randomly (e.g., with a random number table). We then identify and number sequentially all wetlands meeting the specified criteria (e.g., wetland type and size) in each cell sampled. The NWI codes on the map are used to identify wetland type. Wetland size can be measured using a template. For example, we used a transparent grid with 64 cells, with each cell equal to approximately 4 ha. Therefore, a wetland that filled one-fourth of a cell would be 1 ha in size.

The number of grid cells to be sampled was determined by calculating a progressive mean (Marsh 1978). We sampled five cells at a time from the randomly numbered list and calculated the mean number of wetlands per cell meeting our criteria. We repeated this procedure until the mean number of wetlands per cell did not change more than 0.1 and three times the number of wetlands desired were identified. These numbers were chosen to increase the probability that we could obtain a representative sample of the wetlands in the area and have a large enough number of potential sites so that sites eliminated (e.g., because access was denied) could be replaced.

We performed a 5% quality control check on the wetland area measurements taken from the NWI maps, i.e., a second individual repeated 5% of the area measurements, so that we could assess measurement precision. An error level of less than 5% was achieved in our studies. We suggest that the project leader and statistician define meaningful values for the quality control check and the acceptable error level for your particular study.

Finalizing the List of Projects and Natural Wetlands to be Sampled

The activities described above will result in lists of projects and natural wetlands that have, thus far, met the sampling criteria. Replace any sites eliminated from consideration with the next wetland on the appropriate list. Sites are eliminated if: 1) access is denied by the landowner; 2) the wetland project has not been constructed or the natural wetland has been destroyed; 3) a field reconnaissance of the site reveals that it does not meet the specified criteria (e.g., it is the wrong type or size); or 4) conditions on or near the site would be hazardous (e.g., garbage is actively being dumped on the site). We have had to eliminate sites from our lists for all of these reasons. The following discussion describes how to confirm that a site is, indeed, suitable for inclusion in your study.

The first step, if it was not done earlier, is to obtain permission from the landowner to enter the wetland. A contact person is often listed in the permit or project files. Finding the owner of the natural wetlands is more problematic. You can use ad valorem tax maps of the county, visit the site and use detective skills (e.g., talk to adjoining property owners), or contact the owner by mail and then follow up with a phone call. Be prepared for many individuals to deny access to their property. In the Oregon Study, owners denied access for 35% of the natural wetlands from the list of potential sites (Abbruzzese et al. 1988).

After you have eliminated the sites to which access was denied, you must locate each of the remaining sites to ascertain if the wetland projects have been completed and the natural wetlands still exist. Locating the projects and natural wetlands will probably be time-consuming because project files often contain vague information on the locations of the wetland projects, and it is often difficult to locate the natural wetlands from maps. While at the site document the location, ease of access, wetland type, and surrounding land use. Figure 3-7 is an example of a form that could be used. By following the procedures outlined above you will have finalized a list of projects and natural wetlands to be sampled. You also will have compiled information on each site that will be useful as you begin monitoring.

SUMMARY

This chapter presents a strategy for setting sampling priorities and selecting sites for a monitoring program. We recommend that priorities be based on past and projected impacts to wetlands so that efforts will be targeted to locales where the resource is at greatest risk. We also recommend that the ecological setting, in particular the land use surrounding a wetland, be accounted for in site selection. Only when wetland projects are compared with natural wetlands located in a similar setting and exposed to similar ecological conditions, can the performance criteria for a project be within the bounds of what is possible.

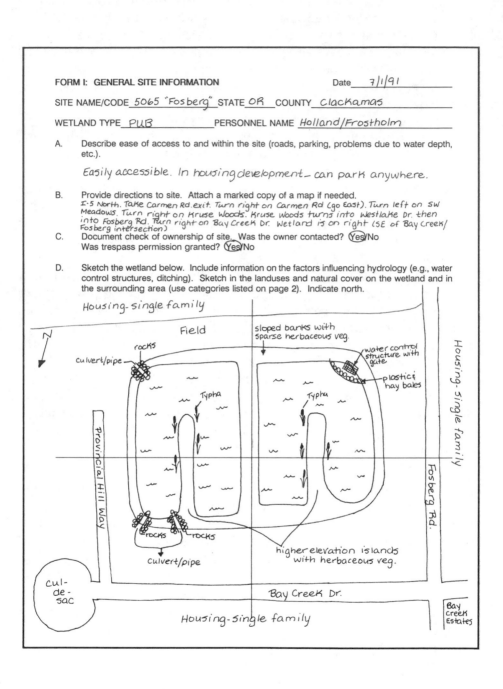

FORM I: GENERAL SITE INFORMATION   Date ___7/1/91___

SITE NAME/CODE _5065 "Fosberg"_ STATE _OR_ COUNTY _Clackamas_

WETLAND TYPE _PUB_   PERSONNEL NAME _Holland/Frostholm_

A.   Describe ease of access to and within the site (roads, parking, problems due to water depth, etc.).

   Easily accessible. In housing development — can park anywhere.

B.   Provide directions to site. Attach a marked copy of a map if needed.
   I-5 North. Take Carmen Rd. exit. Turn right on Carmen Rd (go East). Turn left on SW Meadows. Turn right on Kruse Woods. Kruse Woods turns into Westlake Dr. then into Fosberg Rd. Turn right on Bay Creek Dr. Wetland is on right (SE of Bay Creek/ Fosberg intersection)

C.   Document check of ownership of site. Was the owner contacted? (Yes)/No
   Was trespass permission granted? (Yes)/No

D.   Sketch the wetland below. Include information on the factors influencing hydrology (e.g., water control structures, ditching). Sketch in the landuses and natural cover on the wetland and in the surrounding area (use categories listed on page 2). Indicate north.

Housing-single family

Field

sloped banks with sparse herbaceous veg.

rocks

culvert/pipe

water control structure with gate

Typha   Typha

plastic & hay bales

Housing-single family

N

Provincial Hill Way

rocks   rocks

culvert/pipe

higher elevation islands with herbaceous veg.

Fosberg Rd.

cul-de-sac

Bay Creek Dr.

Bay Creek Estates

Housing-single family

Figure 3-7.   Example of a completed form that can be used during a field reconnaissance to collect information on potential study sites.

An Approach to Improving Decision Making in Wetland Restoration and Creation

38

FORM I: GENERAL SITE INFORMATION   Page 2

I.   Indicate % open water, % vegetated and % non-vegetated areas within the wetland (A-C
      should add up to 100%):

      A. _80_ % open water
                    1. _75_ % unvegetated
                    2. _5_ % with submerged aquatic vegetation

      B. _15_ % vegetated
                    1. _0_ % trees
                    2. _0_ % shrubs (15 feet or less)
                    3. _15_ % herbs

      C. _5_ % unvegetated

TOTAL: 100%

II.   Indicate % relative cover of surrounding areas within 100 meters of the wetland
      boundaries (A-E should add up to 100%):

      A. _10_ % trees
      B. _2_ % shrubs
      C. _8_ % natural herbaceous vegetation
      D. _0_ % water body--specify type:_____
      E. _80_ % human landuse
                    1. _0_ % crops
                    2. _0_ % fallow
                    3. _0_ % grazing
                    4. _0_ % industrial--specify type:_____
                    5. _0_ % commercial
                    6. _5_ % transportation corridor
                    7. _75_ % housing--single family dwellings
                    8. _0_ % housing--multiple family dwellings

TOTAL: 100%   **NOTE: 1-8 should total the percentage value in E.

III. Indicate % of wetland which is disturbed and describe the disturbance (for example,
      ditches, water control structures, dumping, fill, and anything that might be hazardous):

                    5% disturbed - plastic and hay bales
                              3 pipes
                         some stakes left on site

IV. Comments:

              Kingfishers using site.

Figure 3-7. (continued)

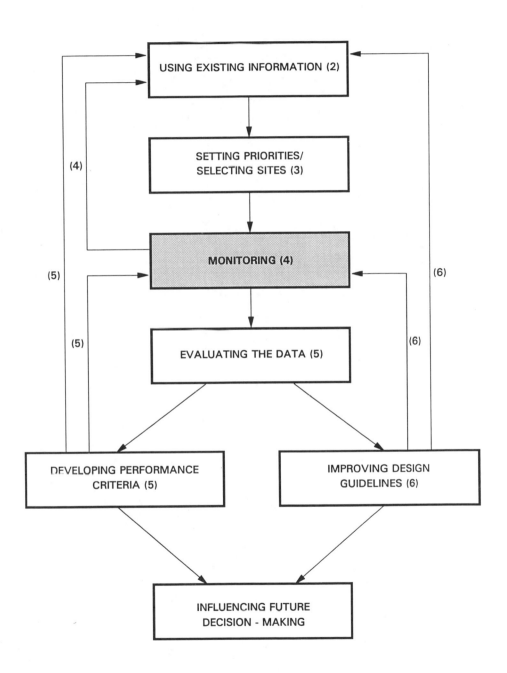

CHAPTER 4

Monitoring Performance

Monitoring is a key element of the WRP Approach. Post-construction monitoring of wetlands, however, is seldom performed (Brooks 1990, Gwin and Kentula 1990, Kusler and Kentula 1990b). Before proposing or initiating a monitoring program it is important to develop a plan based on project objectives as an integral part of the project. The plan documents proposed assessment procedures, timing, and frequency as well as the person or organization responsible. Monitoring decisions should be guided by the type of information that will be needed to determine if project objectives are being met. Who is responsible for performing and overseeing corrections? Will the data from monitoring be used to evaluate compliance with specific permit conditions? Is the purpose of the study to detect an improvement or decline in a wetland's condition or functions? Will the data collected be compatible with information from previous reports or similar studies (Brooks 1990)?

In the past, assessments of wetland projects have emphasized structural rather than functional attributes. Although structural features, such as water levels, are convenient to measure, these features are most useful when related to the functional capability of the wetland. Many structural measures, such as community diversity, become indicators of function when monitored over time. Therefore, it is critical that a performance evaluation of a project consider both functional and structural capabilities (e.g., Marble 1990).

The intensity of post-construction monitoring varies with the environmental significance of the project, the compliance requirements, the age of the project, and the probability of successfully achieving targeted wetland functions. Most wetland projects are designed to provide only a few specific functions. By focusing monitoring efforts on these designated functions, the associated costs can be reduced.

In this chapter we propose some general procedures for performing assessments at three levels of effort. In Chapter 5 we suggest how to analyze the data and use the results. The three assessment levels are: 1) documentation of as-built conditions, 2) routine assessments, and 3) comprehensive assessments. For each of the variables suggested, a brief rationale relating it to wetland function is provided in Table 4-1. The variables and suggested methods for measuring them are presented in Table 4-2 and apply to both projects and natural wetlands. The data collected during each level of assessment are hierarchical. That is, information obtained during an as-built assessment forms the basis for routine and comprehensive assessments that occur later. A hierarchical approach to data collection facilitates making comparative evaluations over time and among similar sites. Ultimately, this process can lead to the development of performance criteria for future wetland projects (Chapter 5).

DOCUMENTATION OF AS-BUILT CONDITIONS

We use the term as-built conditions to refer to actual project conditions at the time of completion. As-built assessment refers to the data collected for evaluation of this condition.

Rationale

First, we recommend checking the wetland for compliance with the design criteria and for agreement with permit conditions or project objectives. Seldom do as-built conditions coincide with original designs. Therefore, it is essential that as-built conditions be documented. At this level of assessment, collect baseline data on project location, morphometry, hydrology, substrate, and vegetation to compare with project objectives and construction plans, and document any differences.

Differences between the design and actual construction may significantly affect the wetland's potential performance. Assessments of as-built conditions help identify noncompliance with permit conditions or project objectives so that corrective steps can be taken as required. Some modifications to the original design are expected due to unforeseen conditions that become evident during construction (Gwin and Kentula 1990). There may also have been mistakes. If construction according to the original design or existing conditions is likely to limit wetland performance, then corrections should be made before completion of the construction phase of the project. These changes should then be justified and accepted before the project is officially approved by the permitting agency or organization responsible for the project.

Table 4-1. Rationale and uses of variables measured in as-built (1), routine (2), and comprehensive (3) assessments of wetland projects and natural wetlands.

VARIABLE	RATIONALE/FUNCTION	SUGGESTED USE(S)
GENERAL		
Location (1)	identifies site on local map	provide baseline map for future assessments
Wetland type (1, 2, 3)	documents project goals (1), successional changes over time (2, 3)	serve as benchmark for future comparisons (1), document expected/appropriate development of the project (2, 3)
Drainage area (1)	determines position in watershed and related functions, flood storage computation	relate to projects functional capability
Surrounding land use (1, 2, 3)	determines inputs to wetland (e.g., nonpoint source pollution, industrial outfalls)	evaluate the need for buffers around wetland, explain changes in wetland performance
MORPHOMETRY		
Area (1, 3)	documents project goals (1), influences habitat value (3), and flood storage (3)	compare to project goals, construction specifications, and future assessments
Slope (1)	influences hydrologic gradient, plant establishment, animal access, characteristics of wetted edge	determine minimum, maximum and mean depths and slopes from topographic profiles for each transect (Figure 4-2)
Perimeter-to-area ratio (1, 3)	influences habitat, edge effect, project goals	determine variation in shape from original design (1), and changes in shape over time (3)
HYDROLOGY		
Water depth (1, 2, 3)	influences flood storage potential, vegetation patterns, wildlife and fisheries habitat	determine hydroperiod, flood storage (Simon et al. 1938), proportion of open water, temporal/seasonal changes
Flow rates (1, 3)	affects wetland characteristics and stability	evaluate water sources, hydrologic modeling
Flow patterns (1, 2, 3)	influences plant establishment and substrate stability and chemistry	serve as benchmark for future assessments of performance

Table 4-1. (continued)

VARIABLE	RATIONALE/FUNCTION	SUGGESTED USE(S)
HYDROLOGY (continued)		
Indirect indicators (1, 2, 3)	provides evidence of hydrology in absence of water during sampling, boundary delineation	establish presence and periodicity of hydrology
SUBSTRATE		
Soil depth (1, 3)	influences suitability as planting and growth medium	verify construction specifications
Soil color (1, 3)	indicates hydric characteristics	determine extent (1) and time of formation of hydric soils (3), boundary delineation
Soil texture (1, 3)	influences suitability as planting and growth medium, root growth and infiltration	verify construction specifications, benchmark for temporal changes
Soil source (1)	provides baseline information	verify construction specifications and identify potential plant propagules
Organic matter (1, 3)	indicates suitability as planting and growth medium, condition of soil processes (Langis and Zedler 1991)	compare to natural wetlands, document temporal changes
Sediment flux (3)	indicates potential for sediment accretion removal, disturbance	measure rates of sediment accretion or erosion for comparisons to natural wetlands, document temporal changes, document/correct erosion
VEGETATION		
Species lists (1, 3)	defines wetland type, habitat, and plant diversity	verify permit or project planting conditions, delineation, calculate weighted averages and ratios (see Chapters 5 and 6)
Coverage (1, 2, 3)	influences use as habitat	verify project goals, benchmark for future assessments
Survivorship (1, 3)	indicates effectiveness of planting methods, influences project goals	evaluate planting success, suggest replanting strategies (3)

Table 4-1. (continued)

VARIABLE	RATIONALE/FUNCTION	SUGGESTED USE(S)
FAUNA		
Observations (1, 2, 3)	indicates use as habitat	evaluate use by common, rare, and exotic species over time
Habitat evaluations (3)	evaluates potential habitat	determine habitat potential over time
Species or community specific sampling (3)	evaluates targeted species or groups of concern	evaluate presence and abundance data over time
WATER QUALITY		
Water samples (1, 3)	indicates water treatment at the site or, disturbance in or around the site	provide baseline data for specific project goals (1), evaluate water treatment function, explain variations in vegetative performance, correlate with faunal use (3)
ADDITIONAL INFORMATION		
Photographic record (1, 2, 3)	provides permanent record for permit file on condition of wetland and surrounding land use	benchmark for temporal assessments, allows for office review of wetland and surrounding buffer
Descriptive narrative (1, 2, 3)	provides additional information and explanation	benchmark for future comparisons

Table 4-2. Methods recommended for measuring variables in as-built (1), routine (2), and comprehensive (3) assessments of wetland projects and natural wetlands.

VARIABLE	AS-BUILT	ROUTINE	COMPREHENSIVE
GENERAL			
Location (1)	use existing map or create map with property boundaries, scale, north arrow, date, latitude and longitude, county and state (add addresses in urban areas and landmarks in rural areas)		
Wetland type (1, 2, 3)	classify intended type(s) (Cowardin et al. 1979)	classify intermediate type(s) (Cowardin et al. 1979)	classify resulting types(s) (Cowardin et al. 1979)
Drainage area (1)	planimeter area from topographic map (ha)		
Surrounding land use (1, 2, 3)	estimate % of surrounding land use, and photograph major types within a minimum of 300 m from the site (Anderson et al. 1976)	estimate % of surrounding land use, and photograph major types within a minimum of 300 m from the site (Anderson et al. 1976)	estimate % of surrounding land use, and photograph major types within a minimum of 300 m from the site (Anderson et al. 1976)
MORPHOMETRY			
Area (1, 3)	determine jurisdictional boundary and use basic survey techniques (Figure 4-1) to create a map of the project (ha)		determine jurisdictional boundary (Federal ICWD 1989) and use basic survey techniques (Figure 4-1) to create a map of the project (ha)
Slope (1)	measure elevation changes at intervals along transects (see Figure 4-2, Gwin and Kentula 1990)		

Table 4-2. (continued)

VARIABLE	AS-BUILT	ROUTINE	COMPREHENSIVE
MORPHOMETRY (continued)			
Perimeter-to-area ratio (1, 3)	planimeter boundary of wetland indicated on the project map and based on jurisdictional boundary (m/ha)		planimeter boundary of wetland indicated on the project map and based on jurisdictional boundary (m/ha)
HYDROLOGY			
Water depth (1, 2, 3)	measure inundation above ground (staff gauge), depth below ground (shallow well, 50-75 mm (2-3") dia. slotted PVC pipe)	measure inundation above ground (staff gauge), depth below ground (shallow well, 50-75 mm (2-3") dia. slotted PVC pipe)	measure inundation above ground (staff gauge), depth below ground (shallow well, 50-75 mm (2-3" dia. slotted PVC pipe)
Flow rates (1, 3)	measure inflow and outflow discharge if present (m^3/s) with flumes or weirs		measure inflow and outflow discharge if present (m^3/s) with flumes or weirs
Flow patterns (1, 2, 3)	use direct observation to indicate major pathways on map	use direct observation to indicate major pathways on map	use direct observation to indicate major pathways on map
Indirect indicators (1, 2, 3)	record observations of indicators (Federal ICWD 1989)	record observations of indicators (Federal ICWD 1989)	record observations of indicators (Federal ICWD 1989)
SUBSTRATE			
Soil depth (1, 3)	use soil auger or dig pit to depth of compacted soil or liner (Federal ICWD 1989)		use soil auger or dig pit to depth of compacted soil or liner (Federal ICWD 1989)
Soil color (1, 3)	use Munsell color chart to determine chroma and hue of matrix and mottles (Federal ICWD 1989)		use Munsell color chart to determine chroma and hue of matrix and mottles (Federal ICWD 1989)

Table 4-2. (continued)

VARIABLE	AS-BUILT	ROUTINE	COMPREHENSIVE
SUBSTRATE (continued)			
Soil texture (1, 3)	use soil texture triangle to classify soil based on feel (Horner and Raedeke 1989)		use soil texture triangle to classify soil based on feel (Horner and Raedeke 1989) or standard methods
Soil source (1)	document source location and addition of any soil amendments (e.g., fertilizer, organic matter, salvaged marsh surface)		
Organic matter (1, 3)	sample during as-built assessment if salvaged marsh surface or other organic materials are added		determine ash-free dry weight from samples (USDA 1984, Blume et al. 1990, NLASI 1983)
Sediment flux (3)	install clay pads at substrate surface as reference points (Cahoon and Turner 1989)		install clay pads at substrate surface as reference points (Cahoon and Turner 1989)
VEGETATION			
Species lists (1, 3)	identify species and wetland indicator and native/introduced status (Reed 1988), document planting locations and methods	identify species and wetland indicator and native/introduced status (Reed 1988)	identify species and wetland indicator and native/introduced status (Reed 1988)
Coverage (1, 2, 3)	estimate cover visually to nearest 10%, map plant communities	estimate cover visually to nearest 10%, map plant communities	estimate cover visually to nearest 10%, map plant communities, collect plot data along transects (Brower and Zar 1984, Leibowitz et al. 1991), collect data for productivity studies
Survivorship (1, 3)	visually determine % of plants alive		visually determine % of plants alive, tag individual shrubs and trees

Table 4-2. (continued)

VARIABLE	AS-BUILT	ROUTINE	COMPREHENSIVE
FAUNA			
Observations (1, 2, 3)	record direct and indirect observations of wildlife, fish and invertebrates	record direct and indirect observations of wildlife, fish and invertebrates	record direct and indirect observations of wildlife, fish and invertebrates
Habitat evaluations (3)	use Habitat Evaluation Procedures (FWS 1980) or comparable method for selected species		use Habitat Evaluation Procedures (FWS 1980) or comparable method for selected species
Species or community specific sampling (3)			select appropriate census techniques (Brooks and Hughes 1988, Brooks et al. 1991, Erwin 1988)
WATER QUALITY			
Water samples (1, 3)	measure appropriate parameters based on project objectives (e.g., pH, conductivity, total suspended solids, nutrients, pollutants)		measure appropriate parameters based on project objectives (e.g., pH, conductivity, total suspended solids, nutrients, pollutants)
ADDITIONAL INFORMATION			
Photographic record (1, 2, 3)	photograph wetland and surrounding landscape from several directions with 50mm lens using 35mm film from permanent photo stations (Horner and Raedeke 1989)	photograph wetland and surrounding landscape from several directions with 50mm lens using 35mm film from permanent photo stations (Horner and Raedeke 1989)	photograph wetland and surrounding landscape from several directions with 50mm lens using 35mm film from permanent photo stations (Horner and Raedeke 1989)
Descriptive narrative (1, 2, 3)	describe and explain notable features and changes for each major variable	describe and explain notable features and changes for each major variable	describe and explain notable features and changes for each major variable

As-built assessments of projects or initial assessments of natural wetlands provide baseline information from which site development and functional performance can be evaluated over time. For example, vegetation data collected during the Oregon Study could not be used to estimate the survival rate of vegetation planted during wetland construction because there was no documentation that planting occurred (Gwin and Kentula 1990). As-built assessments would have provided this information so that vegetation survival rates could have been determined and that aspect of the project design evaluated.

What to Include

The objectives for the documentation of as-built conditions are to collect sufficient information to assess compliance with permit conditions or project objectives and to provide a baseline for future evaluation of project development and performance. The basic elements of an as-built assessment are listed in Tables 4-1 and 4-2. There are both graphic and written components to an as-built assessment. Maps are generated to record wetland area, shape, the patterns of vegetation and open water, major structural components (e.g., water control structures) and surrounding land use (Figures 4-1, 4-2, 4-3). A written narrative augments the graphics and serves as a record of what was done during construction regarding substrate, hydrology, and planting. Additional information may be required. For example, if a project objective is to improve or develop habitat for a specific fish or wildlife species, then an as-built assessment should include a species census, or at least an evaluation of potential habitat at the time of project completion.

The as-built assessment should be completed by the party responsible for construction of the project. In a regulatory situation, this can be ensured by making the assessment a mandatory condition of the permit. Clear guidance should be given as to what is required, how it should be documented, when it should be delivered, and where the documents should be filed. Ideally, as-built assessments will follow immediately upon completion of a project. However, given the variability in scheduling the phases of a wetland project (e.g., design, excavation, planting), as-built assessments may not be completed until months or years after construction. To assist future evaluators, record the approximate time elapsing between completion of the project and the commencement of monitoring. Although the as-built assessment should be conducted at the time the project is completed, delayed data collection is preferable to no assessment at all.

The effort required to produce an as-built description of a site will depend on how closely the construction plans were followed. If the wetland was constructed as planned, very little may be required other than to verify that the plans are correct (e.g., the original site maps will not need to be redrawn). If, however, construction differed from the plans, actual site conditions must be

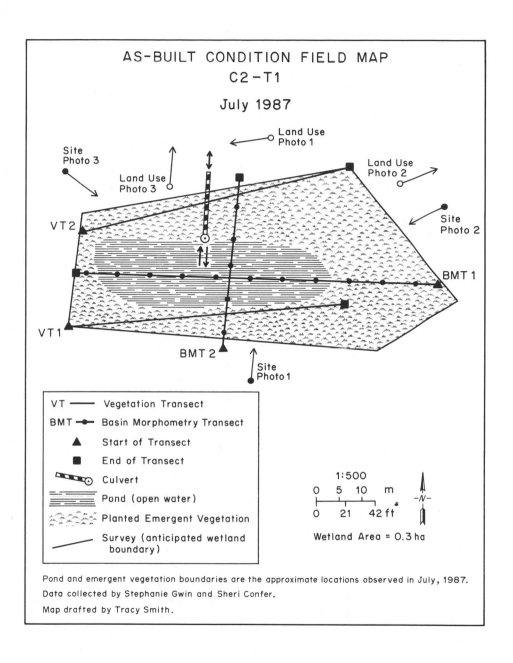

AS-BUILT CONDITION FIELD MAP

C2–T1

July 1987

Land Use Photo 1

Site Photo 3

Land Use Photo 3

Land Use Photo 2

Site Photo 2

VT 2

BMT 1

VT 1

BMT 2

Site Photo 1

VT ———	Vegetation Transect	
BMT ●—●—	Basin Morphometry Transect	
▲	Start of Transect	
■	End of Transect	
	Culvert	
	Pond (open water)	
	Planted Emergent Vegetation	
	Survey (anticipated wetland boundary)	

1:500

0   5   10   m

0   21   42 ft

–N–

Wetland Area = 0.3 ha

Pond and emergent vegetation boundaries are the approximate locations observed in July, 1987.
Data collected by Stephanie Gwin and Sheri Confer.
Map drafted by Tracy Smith.

Figure 4-1. Example of a Field Map to document as-built conditions of a wetland project.

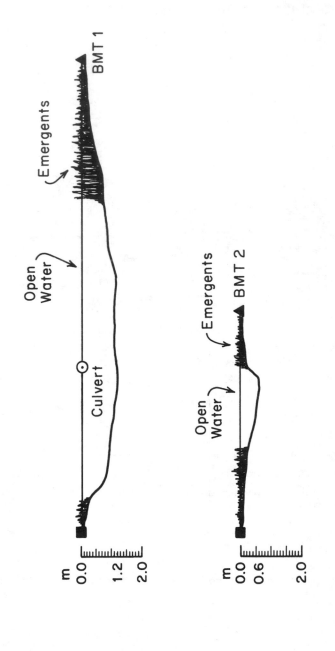

Figure 4-2. Profile of a wetland portraying as-built or current conditions based on elevational measurements from Basin Morphometry Transects (BMT1 & BMT2).

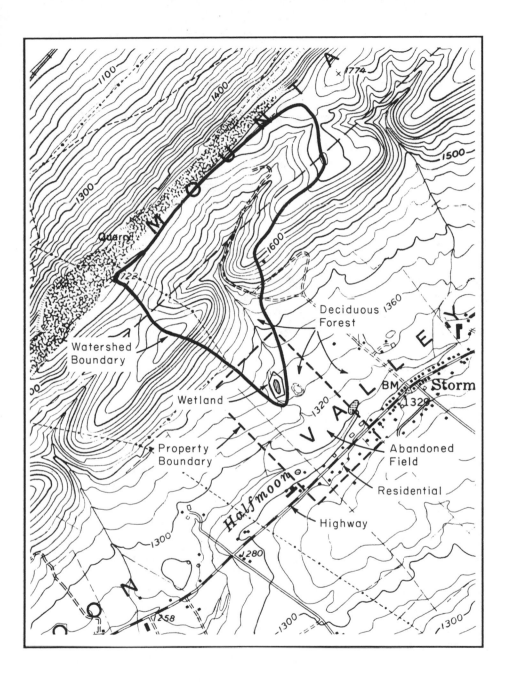

Figure 4-3. Map enlarged from U.S. Geological Survey quadrangle showing drainage area, surrounding land-use, and wetland location.

fully documented. This will require additional mapping and data collection (Tables 4-1 and 4-2).

Once the as-built assessment is complete, differences between what was planned and what was built should be evaluated by the permitting agency or organization responsible for the project. If modifications to the project are necessary at this point, the as-built assessment will need to be updated to reflect these changes. When it is final, file the as-built assessment with the permanent project records so that it is available for comparison with future site assessments.

For natural wetlands, the initial assessment will consist of a base map that documents the conditions found at the site and a supporting narrative. Subsequent assessments will rely upon an accurate portrayal of these initial conditions.

ROUTINE ASSESSMENTS

Routine assessments are simple site examinations or "spot checks" used to monitor and record wetland development. During the check, visual assessments of wetland conditions are compared to maps and photographs from prior visits and the differences noted. This information is used to: 1) identify problems that require correction; 2) provide a record of progress; and 3) determine, in some cases, when site performance warrants releasing the contractor from further responsibility.

Rationale

Routine assessments are less costly and potentially less damaging to the site than more comprehensive assessments because data collection is less intensive and many observations can be made without entering the interior of the wetland (Kadlec 1988). Depending on the size and complexity of the project or the natural wetland being monitored, routine assessments generally take less than a day to complete. They require little equipment and limited training of personnel.

Conduct routine assessments during the first few years after wetland construction when plant communities and hydric soils are becoming established. Comprehensive assessments may be unnecessary or inappropriate until the immediate effects of construction activities have passed. Data from prior studies of natural wetlands or wetland projects can be used to determine if and when routine assessments should be replaced or alternated with more comprehensive assessments. The time required for wetland projects to develop such specific attributes as stable or mature vegetation communities or wildlife habitat can be estimated from these earlier studies.

What to Include

Data collected during routine assessments should reflect project objectives. At minimum, routine assessments include the collection of the types of data noted in Tables 4-1 and 4-2. This set of standard data builds upon data collected during the as-built assessment, as shown in Figure 4-4, and can contribute to a regional database on wetland performance and design (see Chapters 5 and 6).

The decision to collect certain types of data is made on a project-by-project basis, and may result in the use of methods from both routine and comprehensive approaches. For instance, if sediment retention is a stated objective, the routine assessment would include at least a visual inspection of water flow rates and patterns and any associated evidence of sedimentation. Evidence of an alluvial fan at a wetland's inlet may indicate excessive sedimentation. If more accurate sedimentation data are required, then more quantitative methods such as annual measurements of sediment accumulation on feldspar clay pads may be warranted (Cahoon and Turner 1989; Barbara Kleiss, COE, Waterways Experiment Station, Vicksburg, MS, personal communication).

Routine assessments should be repeated at appropriate intervals to determine if the project is on track and objectives are being met, and should be performed during an appropriate time of the year. For instance, if a project objective is flood peak reduction, inspect the site during flood events to see if it receives floodwaters. A comprehensive assessment of flood storage function might involve calculation of the actual volume of water stored (e.g., Simon et al. 1988). Similarly, objectives relating to wildlife use suggest inspecting the site during breeding, nesting, or migration seasons. Inappropriate timing of wetland visits can lead to high variability in the data. Alternatively, high variability in data collected from different routine assessments may indicate that another indicator should be used to assess the wetland function being studied. In addition, some variability will be due to natural changes in the wetland and are to be expected.

Generally, perform routine assessments annually or until you are confident the project is developing as expected. This allows major problems (such as excessive sedimentation or failure of a water control structure) to be identified and corrected expeditiously. Regular annual checks also provide information on the wetland's structural development and functional performance over time. This is essential for determining if performance criteria are being met (see Chapter 5). The information obtained is also needed to establish or evaluate performance criteria for all wetland projects in the region. As this information accumulates in the project record, the frequency and timing of assessments can be modified, as necessary, to produce reliable data.

In summary, annual routine assessments continue until the objectives of the project are met and the contractor is released from contractual obligations,

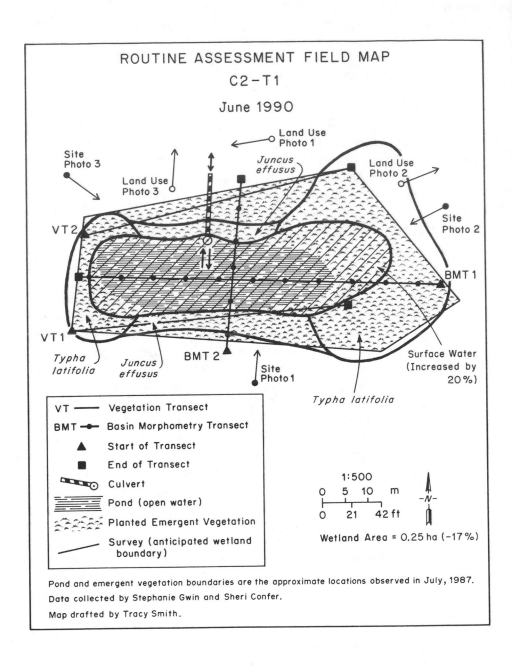

ROUTINE ASSESSMENT FIELD MAP

C2-T1

June 1990

Land Use Photo 1

Site Photo 3

Juncus effusus

Land Use Photo 3

Land Use Photo 2

Site Photo 2

VT2

BMT 1

VT1

Typha latifolia   *Juncus effusus*   BMT 2

Site Photo 1

Surface Water (Increased by 20%)

Typha latifolia

VT ———	Vegetation Transect
BMT ●—●	Basin Morphometry Transect
▲	Start of Transect
■	End of Transect
▨▨○	Culvert
≡≡≡	Pond (open water)
⁓⁓	Planted Emergent Vegetation
╱	Survey (anticipated wetland boundary)

1:500

0   5   10   m

0   21   42 ft

-N-

Wetland Area = 0.25 ha (−17%)

Pond and emergent vegetation boundaries are the approximate locations observed in July, 1987.

Data collected by Stephanie Gwin and Sheri Confer.

Map drafted by Tracy Smith.

Figure 4-4. Example of a Field Map to document conditions found during a routine assessment as compared to the as-built condition. Heavy dark line indicates most recent wetland perimeter and separates areas of dominant vegetation types. Note change in wetland shape as compared to as-built conditions shown on Figure 4-1.

An Approach to Improving Decision Making in Wetland Restoration and Creation

OR until the routine assessments are replaced by more comprehensive monitoring procedures OR until experiences indicate that projects at a certain stage of development need less frequent assessment. Even after contractual obligations are fulfilled, continued periodic, routine checks can provide important information on the wetland's persistence and performance over time.

A report should be compiled after the routine assessment is performed. The report clearly indicates if corrections are required or if more comprehensive monitoring is needed to interpret wetland conditions. The report also documents significant changes at the site that have occurred since the as-built conditions were documented or the last routine assessment was performed.

The completed routine assessment should be furnished to managers at the permitting or sponsoring organization for evaluation. The report can be used by managers to make decisions such as requiring more comprehensive assessments, continuing routine assessments, or releasing the contractor from further responsibility. The routine assessment should be filed with the permanent project records so that it is available for future reference, and appropriate summary information entered into any associated database to keep reports on the project current.

COMPREHENSIVE ASSESSMENTS

Comprehensive assessments generate more complete and quantitative information on the wetland's performance than do routine assessments. Information gathered during comprehensive assessments is important to: 1) identify modifications to the site that are required to meet project objectives; 2) provide a basis for evaluating project design and establishing performance criteria; 3) help explain why a wetland project was or was not successful, and 4) support long-term research efforts.

Rationale

Comprehensive assessments are generally more costly, require more skilled personnel, and can result in greater disturbance to the site, as it is necessary to sample and operate in the interior of the wetland. Therefore, they should not be performed until substrates have stabilized and plant communities are flourishing. The only exception would be to meet the needs of research efforts to evaluate the early development of sites. Comprehensive assessments vary in breadth, detail, and frequency of data collection depending on project objectives and the needs of the sponsoring organization.

Comprehensive assessments should generally be performed when sufficient time has elapsed after wetland construction to allow major wetland characteristics to develop. This may be three to five years for emergent wetlands and longer for forested wetlands. Sometimes, however, specific project conditions require a thorough or partial in-depth assessment at an earlier stage. An example of this is the protection or enhancement of an endangered plant species which requires careful early monitoring of the species' condition throughout several growing seasons. Comprehensive assessments may also be needed if a routine assessment indicates possible problems with the site and additional information is required to determine appropriate corrective actions.

What to Include

There are no specific procedural requirements for comprehensive assessments because the reasons for conducting them vary. Tables 4-1 and 4-2, however, list some possible procedures and why they would be used, and provide additional sources of information on methods.

Data collected during each level of assessment must be compatible with data collected previously. Although data collection and analysis should be of the highest possible quality during all levels of assessment, it is particularly important for the comprehensive assessment because of the effort expended at this level. The rationale used to justify an intensive sampling effort includes specific objectives framed as hypotheses so that defensible conclusions can be reached. In addition, to meet quality assurance objectives, the reasons for collecting data on the chosen set of variables must be carefully thought out and documented. We recommend: 1) development and evaluation of standard operating procedures and sampling protocols by knowledgeable individuals; 2) acknowledgement of possible sources of error and bias in the procedures; and 3) collection and evaluation of quality assurance replicates during all phases of field and laboratory work to maintain scientific defensibility.

Copies of procedures, data, and assessment results should be supplied to the organization responsible for the project. This material is then filed with the permanent project records and becomes available for future site assessments or research to help maintain consistency over time and improve the interpretation of results.

Part of a comprehensive assessment is an analysis and evaluation of the wetland's development and functional performance over time, based on comparisons to as-built conditions and previous routine and comprehensive assessments. The current status of the wetland is determined with respect to intended type and area (as required by the permit conditions or project objectives), and its sustainability as a functional system in the landscape. If design corrections are needed to meet project objectives or to maintain the system, the possible impacts of modifying the existing conditions on the site must also be considered. Alternatively, you may need to reevaluate and adjust the performance criteria required to more realistic levels, given the uncertainty of wetland restoration and creation technology and variations in environmental conditions from year to year.

An Approach to Improving Decision Making in Wetland Restoration and Creation

ASSESSMENT VARIABLES

The variables measured during monitoring at the various levels are discussed in the following sections. We are suggesting methods that we have used successfully in the field.

We realize, however, that other methods may work equally well. See Horner and Raedeke (1989), Adamus and Brandt (1990), and PERL (1990) for additional recommendations.

General Information

Standard information must be collected to identify the location of each project and natural wetland being monitored. The project or permit file will contain much of the required information, but often it will need to be amended during the assessment of as-built conditions to include a narrative description, photographic record, and, most importantly, an accurate map. Because the condition of a wetland often depends upon its surroundings, we recommend determining its position in the watershed (e.g., headwater, stream order, floodplain, isolated), and measuring the receiving drainage area on a topographic map (Figure 4-3). Obtain the watershed boundaries or drainage area from a USGS (1:24,000 scale) quadrangle, unless the wetland is quite small, in which case a survey done in the field may be substituted. Classify both the wetland type and the surrounding land use according to standard systems, such as Cowardin et al. (1979) and Anderson et al. (1976). Then use the map to estimate and record the percentage of each land use type occurring within at least a 300-m band around the wetland (Table 4-2, Figure 4-3). Routine assessments can record observations and changes on the base map created during the as-built assessment (Figure 4-4), to provide consistency over time and reduce the mapping effort during subsequent visits.

Morphometry

Most wetland projects, and many natural wetlands are located in topographic depressions or basins. Measurements of physical features, such as area, slopes, and water depths should be made during all assessments and used to construct a map from an aerial view and topographic profiles from a side view (Figures 4-1 and 4-2). Accurate portrayal of the wetland in its as-built condition is particularly critical, since this will form the basis for future comparisons. These data can be collected along transects as can data on substrate, vegetation, hydrology, and fauna (Figure 4-5).

A site location map should identify watershed position and land use adjacent to the site (Figure 4-3). If accurate as-built maps are available, only limited field work will be needed to complete the assessment. If information on the as-built condition is not available, then field mapping must occur during the first monitoring visit to the site. Use standard wetland delineation proce-

Figure 4-5. Field crew members taking elevation measurements along a transect.

dures to determine the extent of wetland present. Keep in mind that newly constructed wetlands may not have developed all the characteristics necessary to meet the criteria for a jurisdictional wetland. Therefore, it may be necessary to estimate the wetland area based on slopes, planting patterns, and existing hydrology. The map can be modified during future assessments. Although physical morphometry of the wetland is unlikely to change dramatically over time, the jurisdictional boundaries may fluctuate.

Hydrology

Fluctuations of water level and the duration of inundation or saturation determine, in part, the composition of plant communities (Erwin 1988). Inundation of water at the surface can be easily observed and recorded on a map. There are times throughout the year, however, when site visits will not coincide with surface inundation, and when soils are saturated below the surface. Therefore, several shallow wells are commonly installed within a wetland to measure water levels below the surface to depths of 0.5 to 2.0 m, depending on the expected movement of the local water table. Plastic (PVC) pipes 50-75 mm in diameter with narrow, horizontal slots were used successfully in numerous projects. These pipes can also be used to measure depths of surface water when standing water is present, or separate staff gauges can be installed (Horner and Raedeke 1989).

During site visits, describe and record on a map the flow rates and patterns of surface water. For wetlands with distinct inlets or outlets, flumes or weirs can be used to measure discharge. Locate any water control or containment structures on the map and describe them as well. Document and photograph hydrologic indicators as described in the Federal ICWD (1989), e.g., drift lines, water-stained leaves, oxidized root channels. Single monitoring visits during the year are not likely to yield reliable information about wetlands with variable hydroperiods, so we recommend multiple visits to make readings during several seasons.

Substrate

Substrate characteristics often reveal hydric conditions. Characteristics such as soil color and mottling, which indicate the duration and depth of soil saturation, can be determined quickly and require little training for evaluation. Gleyed soils (those predominantly neutral gray in color and occasionally greenish or bluish gray) are typically hydric. Mottle abundance, size, and color usually reflect the duration of the saturation period and indicate whether the soil is hydric (Federal ICWD 1989). Use a Munsell color chart to determine the hue, value, and chroma of both the mottles and the surrounding soil matrix during as-built and comprehensive assessments (Figure 4-6). The percent soil organic matter determines the suitability as a planting and growth

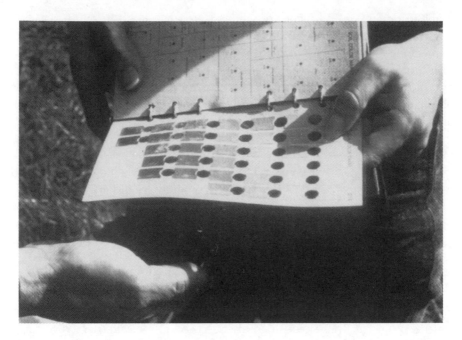

Figure 4-6. Field crew members using a Munsell color chart to determine soil hue, value, and chroma.

medium. The proper percentage of organic matter and the proper soil texture and "hardness" are required to allow penetration by roots and rhizomes (Owen et al. 1989) for vegetation establishment. Soil organic matter also provides necessary nutrients for microbial activity. Refer to Chapter 6 for a more detailed discussion of soil organic matter and substrate characteristics.

Vegetation

A species list and the arrangement of plant communities on the site are commonly used to characterize the vegetation of wetlands. Vegetation data can be collected in a variety of ways (e.g., Brower and Zar 1984, Pielou 1986). For documenting as-built conditions, the identification, coverage, and location of each planted species is essential. These data serve as a bench mark from which to compare the plant community as it matures or changes over time and to determine the survival of plantings. For routine assessments of projects or initial visits to natural wetlands, visual estimation of the percent plant cover and a list of dominant species is usually sufficient. However, comprehensive assessments require a more quantitative approach using quadrats of various sizes (circular, square, or rectangular in shape, 0.1, 0.25, or 1.0 m^2 for herbaceous plants, >1.0 m^2 for shrubs and trees, (Brower and Zar 1984, Horner and Raedeke 1989, Leibowitz et al. 1991) to measure density or coverage (Figure 4-7). The size and shape of quadrats chosen for sampling should be appropri-

Figure 4-7. Botanist reading a vegetation quadrat.

ate for the type of vegetation being sampled, remain constant over time, and be consistent among sites.

Once the extent and boundaries of the plant communities are mapped, and a species list is generated, these data can be used to evaluate wetland dependency (Reed 1988, the appropriate volume; Wentworth et al. 1988) and other characteristics of the plant community (e.g., ratios of native to exotic species; see Chapters 5 and 6). The permit conditions for wetland projects typically focus on the community composition, coverage, and survivorship of plants, so monitoring efforts, are, in part, directed at obtaining these data.

Annual routine assessments generally are performed during the mid-to-late growing season for most wetland types (Brooks 1990), although sampling during other seasons may be more appropriate for certain types (e.g., vernal pools). One factor to consider when choosing a time to sample is that the availability of mature fruits helps in the identification of plant species. In addition, wetland plant communities may be absent or hard to identify at certain times of the year. Check ephemeral wetlands such as vernal pools and some wet meadows when wetland vegetation is present and fruiting, even though hydrologic evidence may be lacking at that time. Although the plant community is usually sampled during the mid-to-late growing season when species composition is of primary concern for the wetland in question, multiple visits may be warranted due to changes throughout the growing season.

Fauna

The habitat value provided to wildlife and fish is frequently cited as a major wetland function and objective of projects. The use of wetlands by diverse faunal communities has influenced both wetland protection and management. Relatively few studies have monitored the diverse fauna that use wetlands (e.g., Brooks and Hughes 1988, Brooks et al. 1991), so sampling protocols and comparative data are relatively scarce as compared to the literature and data on plants. If one objective for a project is to create habitat, then some assessment of habitat condition must be included in the sampling procedures. Minimally, direct and indirect (e.g., tracks, scat) observations of vertebrates and invertebrates must be recorded during all levels of assessment. More quantitative information can be gained by evaluating habitats for a few selected indicator species (e.g., wood frog for northeastern forested wetlands) using the Habitat Evaluation Procedures (HEP) developed by the USFWS (1980). HEP is most applicable for temporal comparisons as the vegetation on the study site matures, or for comparisons between projects and natural wetlands. As required, specific census techniques can be used to determine the presence and abundance of selected faunal groups (Schemnitz 1980, Erwin 1988, Horner and Raedeke 1989, Brooks et al. 1991, Murkin 1984) (Figure 4-8). The timing of censuses can significantly affect results, so we recommend

Figure 4-8. Field crew member collecting invertebrates from an emergence trap.

monitoring the faunal community with careful attention to expected daily and seasonal variations (e.g., early morning surveys for birds and spring breeding surveys for amphibians).

Water Quality

We suggest avoiding a substantial investment in analyzing water samples unless the project of interest has water quality improvement as a primary objective (e.g., constructed to retain nutrients or treat storm water) or a specific pollutant load is expected (e.g., heavy metals or pesticides are found in high concentrations in the adjacent landscape). Implementing a water quality monitoring program for a large population of wetlands may be prohibitively expensive. An additional complication is the inherent spatial and temporal variability found in the chemical characteristics of water in wetlands. The choice of water quality parameters is left to the discretion of the investigator. See Horner and Raedeke (1989) for suggestions on how to implement a water quality monitoring program for wetlands.

Additional Information

Finally, all levels of assessment should include a photographic record and descriptive narrative of conditions present in the wetland and surrounding landscape (Table 4-2). In addition to providing a historical record of visual changes, a photographic record allows you to begin your evaluation of a site in the office. Photographs are no substitute for quantitative data, but are an invaluable aid to documentation of site conditions.

When taking photographs, specify a standard protocol to facilitate intersite and temporal comparisons. We suggest using a 35-mm camera and 50-mm lens. Color film is recommended; prints (and the accompanying negatives) provide a more convenient format for a permit file, but slides are also useful. Take sufficient numbers of photographs to allow evaluation of the site from all directions. Consider taking photographs from a permanent station along the four major points of a compass. Indicate the locations of photo stations and the directions of photos on the site map.

A brief narrative should be included to describe features or findings that do not fit into the above categories (e.g., the water control structure was vandalized resulting in a reduction of water depth), and to document the current condition or observed changes from past assessments (e.g., the water control structure should be repaired because the drop in water level has decreased the wetland area by 20%).

Developing an Efficient Sampling Strategy

The cost of post-construction assessments varies dramatically with the methods and intensity of data collection. Given the potentially large number of wetlands in a target population, this cost must be balanced against the value of the information collected. Devoting a high level of effort and expense to data collection is neither appropriate nor necessary for all assessments. To reduce costs and increase accuracy, select cost effective and efficient assessment methods (e.g., PERL 1990).

One way to increase assessment efficiency is to use compatible field methods, units, data analysis, and reporting procedures. This promotes more accurate, cost effective, and meaningful post-construction assessments. A standardized data form designed for ease of data entry facilitates development of consistent assessment procedures and allows the aggregation of data from many wetlands, which in turn, enables the development of a regional wetland database.

Establishing permanent sampling plots is another way to make different assessment levels compatible. Visits to permanent plots are not likely to significantly affect the wetland for any of the three levels of assessment proposed. If permanent plots are used, however, sampling must be nondestructive and using various routes to access the plots may be adviseable to avoid creating paths. For example, plant specimens should not be removed from plots unless absolutely necessary to avoid introducing a possible bias in species occurrence or abundance.

Agency personnel typically must evaluate multiple sites within a short time period. A schedule designed to allow sampling of sites in geographic proximity will reduce the time spent traveling to sites, thus improving sampling efficiency and reducing costs.

Another method for reducing cost is to use local volunteers. If properly trained and supervised, volunteers can be a source of high quality assistance. The special insert following this chapter discusses the role of local volunteers in site assessments.

The sampling strategy determines when, where, and how to collect data. Your strategy will vary with the project goals and the specific assessment level being performed, but the sampling methods should remain consistent. This promotes data comparability among sequential assessments. For instance, vegetation assessments should be performed in plant communities during the same part of the growing season each year rather than at the same date because of differences in weather year to year. The ability to compare results of different assessments over time enhances your ability to evaluate performance criteria as wetland restoration and creation technology improves.

Data Quality

It is important that the field methods selected provide high quality data that are scientifically defensible. Variation in data due to sampling, collection, and processing methods must be as low as possible, or actual changes in site conditions may not be detected. In our studies, we have considered five basic quality assurance components: precision, accuracy, completeness, representativeness, and comparability. Each component addresses a different aspect of data quality.

Precision is a measure of mutual agreement among individual measurements of the same variable, usually under prescribed similar conditions (Sherman et al. 1991, adapted from Verner, 1990). Precision is usually expressed in terms of the standard deviation, however, precision is calculated differently depending on the variable and method used for measurement. You can use field and laboratory duplicates, standard procedures, and process repetition by separate individuals to achieve better precision in your data.

Accuracy is the degree to which a measurement represents the true or accepted reference value of the variable measured (Sherman et al. 1991). Accuracy depends on the technique used to measure the variable and the care with which it is executed. It is difficult to assess accuracy for many field measurements, however, examples of ways to improve the accuracy of your data include: use of tested standard procedures, training of field crews, and use of standard reference materials.

Completeness is a measure of the amount of valid data obtained compared to the amount that was expected under ideal circumstances (Sherman et al.

1991). You may not be able to collect all the expected data due to time constraints, adverse field conditions, or sample and data loss. If the number of
samples taken is less than that originally intended, seek the advice of a statistician to make sure that the sample size is large enough to produce data that adequately represent site conditions. Proper sample and data handling procedures (e.g., labeled samples and legible data forms) can reduce the chance of
lost information.

Representativeness expresses the degree to which data accurately and pre-
cisely represent a characteristic of the variable of interest (Sherman et al.
1991). Consider representativeness during site selection to ensure that the
sites chosen are representative of the population (See Chapter 3). Transect and
plot establishment should also represent typical conditions of the site being
sampled.

Comparability expresses the confidence with which one data set can be
compared to another (Sherman et al. 1991). By comparing duplicate data collected by different field personnel working on the same sites and plots, an estimate of data variability caused by individual bias can be obtained. If the variability of data collected by different individuals exceeds the inherent
variability of the measurements of the same variable, then the sampling strategy will require modification. Field crew training and standardized procedures
will improve data comparability.

Where to Collect Samples

Sampling strategy depends on the variables being measured, their distribution across the site, and the intended use of the data. For instance, water samples collected to provide information on how a wetland functions as a nutrient
sink need only be collected at water inlets and outlets.

Wetlands frequently have heterogeneous distributions of vegetation and
soils. These distributions can be sampled with a systematic (i.e., samples are
taken at predetermined intervals) or stratified random (i.e., samples are taken
randomly within subdivisions of the unit being sampled) sampling design.
Transects for systematic sampling should be established parallel to environmental gradients such as moisture or elevation. If natural variation is not associated with an identifiable environmental gradient, collect samples randomly
or systematically from each major stratum so that the full range of the
variable's attributes is represented in the data.

How Many Samples to Collect

Both the cost of collecting data and the potential site damage due to trampling during field work increase with sampling intensity. Sampling intensity,
however, must be high enough to produce data that adequately represent site
conditions. High variability within the site increases the number of samples

required. The tradeoff is cost and site damage versus data precision. Krebs (1989) is a good reference for various methods of determining the optimum number of samples to collect for ecological studies, as are most biometrics or statistical reference books (e.g., Snedecor and Cochran 1980).

When to Collect Samples

The assessments discussed in this chapter serve different purposes and thus, the timing and frequency of sample collection will vary. In general, sampling should be timed to match important phenomena relating to the project objectives (Brooks and Hughes 1988, White et al. 1990, PERL 1990, Leibowitz et al. 1991). As noted, wildlife habitat use should be checked during the time the target species is likely to occupy the wetland, and vegetation should be checked during the part of the growing season when plants can be most easily identified. Hydrologic sampling in a region with evenly distributed patterns of precipitation throughout the year will differ from that of an arid region or one with distinct wet and dry seasons.

Sampling frequency required depends on the preconstruction conditions at the project site (White et al. 1990). The probability of successfully restoring a wetland often depends on the extent to which the wetland is degraded. If most of the attributes of a functioning wetland (e.g., hydric soil, wetland vegetation, and correct hydrology) are still present or easily repaired, the project is more likely to succeed. On the other hand, if major wetland attributes have been destroyed, or if a wetland is being created on an upland site, the success of the project becomes more uncertain. Projects with very uncertain outcomes require more frequent and/or intensive monitoring so that timely design corrections can be made if necessary (White et al. 1990).

Controlling Damage to the Site

Performing site assessments can damage a developing wetland. Therefore, post-construction assessments should be designed to minimize activities within the wetland. It is best to minimize walking within the site during reconnaissance and sampling activities. Approach permanent sampling transects or points by alternate routes during successive visits. Do not traverse the same place repeatedly to avoid developing a trail. Trails, besides destroying vegetation and altering water flow, are often invitations for other people to enter the wetland, sometimes on horseback, motorcycles or off-road vehicles.

SUMMARY

Insight into the probable success of wetland restoration and creation efforts will enable you to make the sound wetland management and policy decisions required to protect the resource. Post-construction wetland assessments are essential to confirm compliance with permit conditions or project objectives and to ensure that projects provide the functions expected.

The assessment procedures used in our Approach are selected to enable incorporation of the data into a regional database that characterizes both wetland projects and natural wetlands. The database will provide a basis for refining regional wetland performance criteria and design guidelines and identifying design characteristics that are most likely to produce the desired functional results. In addition, recording performance levels of wetland projects promotes the establishment of appropriate and attainable wetland performance criteria.

Assessment procedures should produce high quality, cost effective data. Adequate information must be collected to assess and promote project success, and to help refine regional mitigation policy. At the same time, assessment activity costs must be minimized to reduce the financial burden on both the public and private sectors.

Using the WRP Approach, a monitoring plan based on project objectives is established as an integral part of the wetland project. Assessment procedures, and timing and frequency of sampling are clearly documented. The plan also identifies the person or organization responsible for performing the assessments, who is to receive and archive the reports, and who is responsible for performing and overseeing corrections. Because corrections can be costly, a strategy for evaluating the assessment findings is included as part of the plan.

To effectively protect wetland resources, we must identify which wetland types and functions can reliably be replaced in a given time and geographic region. We need information on how to manage wetlands, on which wetland designs work, and on how restored and created wetlands are likely to persist and function over time. A well planned monitoring program plays an essential role in acquiring this knowledge.

Volunteers and Natural Resource Monitoring

by Neal Maine

Editor's Note: We are pleased to offer the following special section as an example of how a monitoring program can be implemented by an agency despite limited staff and funding. Author, Neal Maine, exemplifies the positive impact one person can have on protecting coastal resources. He won the prestigious Chevron Conservation Award in 1988 after being chosen to represent Oregon by then Governer Neil Goldschmidt because "he educates Oregonians about our coastal resources by encouraging citizen involvement in conservation projects". Neal's contribution spans 26 years as a teacher in the Seaside Schools and as an instigator and volunteer in north coast wetland conservation projects. Asked how he has managed to achieve so much, Neal says: "When I see a need or where something is being threatened and nobody is doing anything about it, I just try".

Volunteers and Environmental Protection

Protection of the environment directly affects quality of life values. For this reason, environmental issues are often the focus of community action. Whether it is a group effort to preserve habitat for a valued species or to clean up a polluted area, many people are becoming more involved in protecting and restoring their environment. Federal and state agencies and nonprofit organizations promote involvement by volunteers in environmental activities because they know how essential local support is to the ongoing success of any project. Local participation includes activities such as community education, helping with work projects, building trails, recording observations, and "adopting" resources as varied as trees, streams and animals.

- In the state of Washington, the Adopt-A-Stream Foundation invites volunteers to adopt a stream or wetland, collect data about the resource, and report the data to a planning organization for use in making decisions about needs for resource management.

- Volunteers work with the Oregon Department of Fish and Wildlife to conduct salmon spawn surveys, rehabilitate streams, and incubate salmon in streamside hatchboxes to help sustain Northwest salmon as both food and game fish.

- The National Wildlife Federation encourages individuals and groups to assist with winter bald eagle counts to ensure that the eagle population is not diminishing.

- The Nature Conservancy invites individuals and groups to participate in stewardship programs at selected reserves.

- Cannon Beach, Oregon, community volunteers educate visitors to the Haystack Rock tidal pools so they can enjoy the area without harming the pools or their inhabitants.

- The Environmental Protection Agency (EPA) has just completed a program called Streamwalk in Region 10 in which volunteers work with scientists to collect general information about a stream by mapping surveyed areas, characterizing the stream, and determining the cause of any adverse conditions found in the stream.

All of these programs involve citizens in the process of monitoring and protecting natural resources. When local citizens become involved with natural resources and their management, a new kind of protection emerges. The awareness of the entire community is focused on the problem at hand, *the* environment becomes *our* environment and is afforded the respect and protection reserved for the Earth as our shared home.

MONITORING MITIGATION PROJECTS

Funding for wetland mitigation, under Section 404 of the Clean Water Act, is based on the need to meet the legal requirements of proposals by developers, agencies, municipalities, and in some cases, individuals, which involve wetland alteration. Support for on-site monitoring of created or restored wetlands is often minimal. It is usually limited to single annual visits to the site for subjective evaluation. The collection of quantitative data is seldom required, although resource managers know that if they are to answer broad questions about the success of mitigation in restoring wetlands or specific questions about individual project success, the collection of valid data must be an integral part of all projects. A major limitation for including detailed on-site environmental monitoring in mitigation plans is the cost of professional services for collecting field data. Logistic problems such as transportation and scheduling constraints can also limit the collection of data on daily, weekly or monthly cycles. This is particularly true when sites to be monitored are some distance from qualified scientific staff. Providing opportunities for participation by volunteers can account for significant cost savings in professional services and travel expenses. More important, while helping collect specific data for a research project, volunteers begin to understand research goals and gain insight into the biological processes that occur in both created and natural wetlands. A local group of educated wetland supporters is created along with the wetland.

Aldo Leopold addressed this type of research 50 years ago in an essay in the Sand County Almanac on the development of the land ethic.   He stated, "The last decade, for example, has disclosed a totally new form of sport, which does not destroy wildlife, which uses gadgets without being used by them, which outflanks the problem of posted land, and which greatly increases the human carrying capacity of a unit area. This sport knows no bag limit, no closed season. It needs teachers, but not wardens. It calls for a new woodcraft of the highest cultural value. The sport I refer to is wildlife (resource) research." Leopold concluded that, "The more difficult and laborious research problems must doubtless remain in professional hands, but there are plenty of problems suitable for all grades of amateurs."

The following sections discuss the concept of supporting an increased level of community involvement in scientific research activities, one in which local volunteers become an integral part of the effort. The discussion centers around how to build a research team that includes volunteers, using as an example the Trail's End wetland creation project in which six amateurs became part of a wetlands research team in Seaside, Oregon. Wetland research is the focus of this particular volunteer project, but local volunteers can be successful coworkers in most monitoring efforts.

THE TRAIL'S END PROJECT

Comparing monitoring data from natural wetlands and wetland projects enables scientists to set standards for restored or created wetlands based on the quality that is attainable in that region and ecological setting. For example, scientists can identify when developing plant communities do not match the natural assemblages, and recommend corrections to project design based on typical local wetlands. The Trail's End study, conducted in the spring of 1989 in northwest Oregon, was designed to use volunteers to collect extensive data for a detailed monitoring study of a 15 acre created wetland, Trail's End, and three natural marshes. One of the objectives of the project was to test the use of volunteers for collecting scientific data within the context of a wetlands re-search project. Specifically, the objective directed researchers to evaluate the use of citizen volunteers in many aspects of the data collection.

Six volunteers from the Seaside community were recruited and trained to work with scientists on contract to the Environmental Protection Agency. These local volunteers accepted the responsibility for collecting data on vege-tation, water chemistry, emergent and benthic invertebrates, hydrology, soils, and bird use. They also participated in mapping the sites, developing a photo record, and testing methods.

We learned from the Trail's End experience that the process of putting to-gether the local team should receive special attention. This team is the spon-soring organization's link with the data collection activities and the wetland re-source. The final accuracy of the data, and therefore, the reliability of the data, depend on the effectiveness of the initial team-building and training ef-fort.

TEAM BUILDING

Individuals who choose to become volunteers are the key to the success of a project. They are asked to put in many hours of work, often for long periods of time on a given day, and to collect data of a quality that will meet the stan-dards of the sponsoring organization. In many cases, individuals are already involved with issues, activities, or management related to wetlands and wet-land creation. These individuals are the best candidates for volunteers and for the role of Local Team Leader.

From the initial contact with the local community through the close of the project, every effort should be made to create a cooperative working environ-ment in which volunteers and scientists are part of a real team effort. To do this, sharing of information from goals and objectives to the final evaluation of the project and possible use of the results is essential. Volunteers can be in-volved in planning their part of the process, or in making decisions about pos-sible changes; they often have creative suggestions and superior knowledge of local resources.

We used a model with four interrelated roles in which volunteers and sci-entists work cooperatively. Each role demands certain responsibilities and de-sired attitudes and skills:

PRINCIPAL INVESTIGATOR

- Represents the sponsoring organization

- Responsible for the project's success

- Proven expertise in the design, execution, and management of major research projects

- Thorough knowledge of the specific wetlands under study

- Commitment to the concept of using volunteers in a research project

- Good communication and interpersonal skills

OFF-SITE TEAM LEADER

- Represents the Principal Investigator

- Liaison between the scientists and the volunteers

- Responsible for on-site procedural decisions

- Thorough knowledge of wetland ecology

- Experience in identifying wetland flora and fauna

- Good communication and interpersonal skills

LOCAL TEAM LEADER

- Represents the volunteers

- Liaison between the Principal Investigator and the community

- (CRITICAL) Experience in organizing volunteers for research project

- Leadership skills for organizing and maintaining a productive and cooperative team

- Commitment to the success of the project and eagerness to learn

- Good communication and interpersonal skills for working with local agencies and the community

- Sound understanding of basic ecological concepts

- Knowledge of wetland flora and fauna

Volunteers

- Should be from the local area

- Willing to make a major commitment to the project

- Interest in protecting wetlands, plants and wildlife

- Some knowledge of natural history and science

- Eagerness to learn how to perform assigned tasks

- Ability to attend to detail and adhere to data collection schedules

- Ability to participate cooperatively in a team effort

The Principal Investigator and the Off-Site Team Leader are chosen by the sponsoring organization. The Local Team Leader and the Volunteers must be chosen with concern for both commitment and ability.

Local Team Leader

Following the designation of the Local Team Leader by the Principal Investigator and the Off-site Team Leader, a briefing meeting should be held to give the on-site person a full sense of the project scope and the general nature of the methods. It is critical at this time to establish the levels of responsibility for the sponsoring agency, the Local Team Leader and the volunteers. The Local Team Leader, who must be knowledgeable about community resources, will make personal contacts to recruit volunteers. This individual's clear understanding of his/her role in providing leadership by coordinating the data collection effort is essential for the success of the volunteer effort. It is also essential for the eventual validation of the field data collected and therefore, the reliability of the project results. An important factor in the success of the Trail's End project was the Local Team Leader's skillful leadership, planning and organization in determining the responsibilities of the volunteers and supporting them in their efforts.

Volunteers

The Local Team Leader contacts individuals who might be interested in volunteering for a community-based research project. It is helpful if they have had some personal association with research, although not necessarily as a professional scientist. These individuals are often found teaching science in schools, studying science in college, and volunteering in natural history programs. Others have had previous experience in a science-related profession, but are now working in different careers, and some have never had professional or academic training in science, yet have developed their skills through personal interest. The most critical attributes of successful volunteers are commitment to the success of the project; ability to attend to detail, adhere to data collection schedules, and work cooperatively; and eagerness to learn.

The Local Team Leader develops a list of the prospective volunteers. After the scope of study has been clearly defined, the Off-site Team Leader works with the Local Team Leader to match the backgrounds and skills of the volunteers to the specific study tasks. Personal contact is generally the most effective way to make the final decisions. An open meeting in the community stimulates interest and gives both team leaders a chance to meet with all of the potential volunteers. The meeting should be structured so that there is time for volunteers to introduce themselves, share their reasons for wanting to participate in the project, and work together in planning the next meeting around their needs for formal instruction.

Both initial and closing interviews are an integral part of evaluating use of local volunteers in data collection. Interviews with the research volunteers who participated in the Trail's End project revealed that they were interested in participating because of the chance to be a part of doing science, not studying about it; a personal interest in wetlands and natural history; the opportunity to use their skills and knowledge in a wetland setting; personal contact with professional researchers; and the excitement of being part of a study on local natural resources. By the close of the project, these volunteers were also concerned that without ongoing monitoring, local wetland mitigation projects would not meet the public need for protecting wetland values. Three of the volunteers at Trail's End were science teachers who transferred their skills to their high school students. The students are continuing to collect long-term data using the field techniques from the original study.

TRAINING FOR VOLUNTEERS

After the team is identified, the next step is to complete the training that prepares the volunteers and the Local Team Leader to collect data at a level that meets the standards for the study established by the Principal Investigator. The data not only have to be accurate, they also must meet the Quality Assurance (QA) standards of the sponsoring agency (EPA in the case of Trail's End).

The introductory training session is the first occasion on which all participants meet and is a good opportunity to begin building the team concept into the project. Encouraging everyone to contribute ideas and ask questions, and making decisions by consensus when appropriate, are two techniques for letting the volunteers know they are important to the project's success. In the first phase of the Trail's End project, presentations by scientists on the research objectives of the project, wetland ecology, data collection methods, correct use of reporting forms, and a discussion of the components of QA helped volunteers understand both the broad view of the project and the specific needs of the sponsoring agency. Training sessions were informal, with lots of time for volunteers to ask questions about the project and to clarify any personal concerns about participation.

Following initial training sessions, participants should be taken to the field where they can get hands-on experience with each of the procedures to be used in the project. For the Trail's End project, EPA staff presented instruction about a task, then volunteers performed the task while scientists played the role of QA observer. For example, using protocol instructions developed for this study, volunteers conducted water quality tests for oxygen and pH. The objectives were to compare water quality at the created wetland with the three natural wetlands and to test the applicability of the chemical field kits. Results were used to check the comparability of data between individuals and replicates. Each volunteer had an opportunity to work through the procedure and record data on the appropriate form. They received immediate feedback on the accuracy of the data. The training staff was sensitive to the concerns of the volunteers, who were allowed to repeat a technique until they were confident in its use. All volunteers agreed that this was the most critical part of the training. The hands-on experience enabled them to master the techniques and understand the expectations of the EPA staff. Volunteers also had the opportunity to change roles, from data collector to QA observer.

In some cases, such as the water quality sampling, data collection depended entirely on the skills and knowledge of the volunteers. In others, the volunteers supported the scientific staff by recording data or helping with equipment. After completing training, volunteers selected areas for which they would be accountable for data collection during the study. They then worked with the Off-site Team Leader and the Local Team Leader to establish a monthly schedule with a built-in QA cycle. As volunteers collected their data, the completed data sheets were given to the Local Team Leader to copy and file, with the originals going to the Off-site Team Leader for permanent project records.

SUPPORT FOR VOLUNTEERS

Ensuring proper training and a positive atmosphere for the volunteers helps ensure quality data collection. The most effective way to support volunteers is to keep them informed about the project and its progress so that they know why they are doing what they are doing and why it is important. Volunteers will devote much time and energy to a project when they feel that they are part of the team. Creating opportunities for local involvement in the study of wetlands resources and their management sends a clear message to the volunteers that public agencies are accessible and local involvement is important. Although they have less specialized knowledge and skill than their professional counterparts, volunteers feel just as strongly about protecting the environment. Many citizens believe that natural resources are truly a part of the public trust and as opportunities open up for cooperative ventures, they are ready to work to protect resource values for all to share. The Trail's End project combined volunteers and scientists in an environmental partnership, and both groups said, "It worked great!"

Volunteers should receive feedback early in the project on how they are doing. Immediate feedback about problems in data recording, questions on technique, or concerns about procedures helps relieve anxiety about data collection problems and helps prevent the possibility of lost data caused by improper use of procedures. During training, it should be emphasized that volunteers are making important contributions to the study and that they are free to telephone the team leaders about any concerns.

If possible, volunteers should be given updates on the data they are collecting, including any preliminary trends, patterns, or new information. Although resource scientists usually wait until data collection is complete to conduct analyses, volunteers appreciate preliminary observations, even though limited, so that they can see the results of their work. Volunteers should not be isolated after they have started work on a particular data collection task. They need to contact each other and share information on experiences and topics of mutual interest. The Local Team Leader can play an important role in maintaining this communication by sending brief newsletters or notes to the volunteers. Any publicity about the project (newspaper, television, newsletters, etc.) should be brought to the attention of the volunteers before it becomes general public knowledge. Weekly meetings during peak data collection times can also keep everyone informed. A debriefing session midway through the project and again near the end help meet this need. At the close of the project, volunteers need to know how project results will be used and why they are important in a larger context. The Off-Site Team Leader and the Principal Investigator can help keep volunteers informed by sending them preprints of journal articles resulting from the project.

Some unstructured social time is important. Volunteers need to have some free time with each other and with resource personnel. Lunches and planned break times give volunteers opportunities for extended conversations with staff on topics of common interest, in this case, wetlands. One major social event, such as a picnic away from the project, with family members included, can add to the overall experience of everyone involved in the project. For the Trail's End Project, a salmon barbecue at the home of the On-site Team Leader provided an excellent setting for informal sharing. The research sponsor needs to accomplish the project tasks, meet deadlines, consider costs, and maintain the highest level of scientific standards, but personal exchange and a shared awareness of the importance of natural systems at work do not have to be sacrificed in the process.

Opening research projects to local volunteers not only aids in data collection and other tasks, it heightens awareness in the community and among resource staff that protecting resources is everyone's job. It is important to remember that many people interested in becoming volunteers are working on other resource projects that may be of interest to the scientific staff. Their participation can broaden the sensitivity of the resource specialists to the interrelated nature of community-based projects.

BENEFITS FOR VOLUNTEERS AND RESOURCE MANAGERS

Local citizens often believe that some distant person is in charge of their resources. Typically, resource managers come to town, do what they do, and leave. There may be some information in the local newspaper about their work, but more likely there will not be unless major issues arise, such as health, natural resource conservation, contested development, or infringement on local government. It is unfortunate that resource monitoring and protection often remain unnoticed. Inviting interested local citizens to become part of a study helps establish a line of communication with resource organizations. It gives interested people an opportunity to contribute to the management process. Most of all, it gives the volunteers a chance to join in the excitement of conducting research and seeing the results used for making better resource management decisions. In the case of Trail's End, it was satisfying and rewarding when the data the volunteers collected passed the EPA review process. The scientists and volunteers had gained open communication, quality data, and ongoing community involvement in the protection of local wetlands.

Involving local citizens at the research level creates a small, but informed group of citizens who are sensitive to the protection of natural resources at a more sophisticated level than the general public. Even though the project ends, and the resource professionals go on to other problems, the new knowledge and awareness, in this case about wetlands and natural history, stays in the community with the volunteers. Local volunteers who helped with the Trail's End project have continued to collect data using the techniques learned during the project. They also added a bird banding element to the study because they wanted to know more about the migratory birds using the wetlands.

Individuals who become involved in a project of this type will begin to influence their friends, coworkers, students, and local governments.

According to their own evaluations made at the close of the project, the individuals who participated in the Trail's End study found the experience and the learning more than worth the effort. The volunteers were doing what they like to do best: being outdoors, protecting natural resources, and sharing that experience with others of similar interests. Volunteers left the project feeling that they had made a positive contribution, enjoyed the opportunity to work with dedicated researchers, become part of a local team, and contributed to wetlands protection. Many also renewed their connection with the earth. As one Trail's End volunteer put it, "The wetland became my friend. I kept going back to say 'hi' to the yellowlegs".

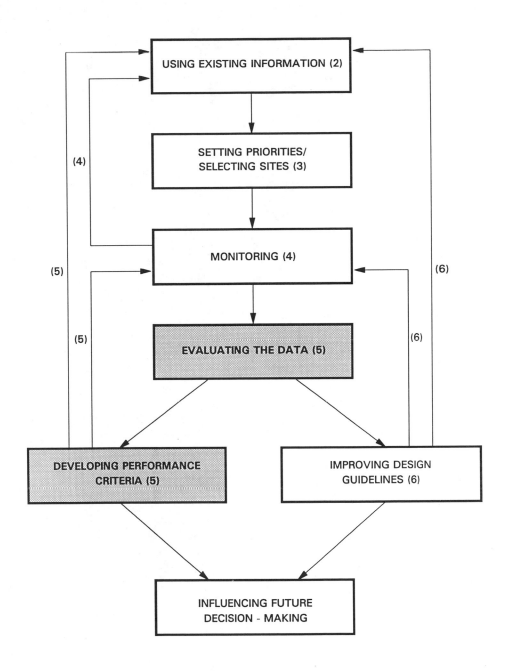

CHAPTER 5

Evaluating the Data and Developing Performance Criteria

The monitoring reports and other data collected for wetland projects are kept on file by many state and federal agencies. However, these reports are rarely used. Often the reports are reviewed only to document that the wetlands were monitored as required (Quammen 1986). In this chapter we demonstrate methods a manager can use to organize incoming data to track the progress of projects and to develop criteria for the evaluation of future projects. Specifically, we discuss ways to represent and evaluate the data, and ways to use the data to set performance criteria. Using the data to compare past and present projects and natural wetlands can help managers set goals, anticipate future problems, and plan for the long-term success of wetland projects.

SUGGESTED WAYS TO REPRESENT THE DATA COLLECTED

We use four different graphs to display the monitoring data: performance curves, summary or descriptive graphs, time series graphs, and characterization curves. These graphs differ in the amount(s) and type(s) of information that can be obtained, and the intensity of data collection necessary. However, all can be used to compare projects and natural wetlands and, therefore, to set criteria for the evaluation of future projects. In the following sections we describe the graphs and curves and present examples of how to create and use them to set performance criteria. Most of the discussion is given to the performance curves and the summary or descriptive graphs, because they were the ones we used most often.

Performance Curves

The hypothetical performance curve described in Chapter 1 displays the changes in a function in wetland projects over time as compared to similar natural wetlands. As discussed in Chapter 3, the curve can be generated in two different ways depending on how you sample. One approach is to follow the development of similar aged projects by sampling the same projects and natural wetlands over time (Figure 3-1a). The other is to gather data from the projects and natural wetlands at one time, but to document development by sampling projects representing a range of ages (Figure 3-1b). The latter is how we generated the performance curves from the results of our field studies (Figure 5-1). Because the age of the natural wetlands is not usually known, the values of the mean and standard error for the measurement of function are placed on the appropriate location on the y-axis.

Figure 5-1 shows that most of the projects had a lower level of percent organic matter than the natural wetlands, suggesting the pattern in the hypothetical examples (Figure 1-2). In reality the curve could take a variety of shapes. It is possible that both recently constructed and mature wetland projects could have a level of a particular function that is higher, lower, or the same as natural wetlands. In addition, the pattern of development of projects could be expressed by a linear, quadratic, logistic, or some other relationship (Figure 5-2).

Although we have a limited amount of information on the development of projects, two things are beginning to be evident. First, the shape of the performance curve will probably vary with wetland type and function. Second, we believe that the pattern of development may be similar for the same wetland type and function in different areas of the country. For example, we calculated a plant species diversity index for each of the created and natural wetlands from the Connecticut, Florida, and Oregon Studies. Even though the means for the created and natural wetlands varied between the states, in general, the diversity of plants found on the created wetlands was initially greater than or equal to that found on the natural wetlands (Figure 5-3).

There are many uses for performance curves, including evaluation and comparison of projects sampled over time (Figure 3-1a), and evaluation and comparison of projects at one time (Figure 3-1b). In addition, how frequently to monitor the projects can be determined from the curves by noting yearly changes in variables of interest. If, for example, the percent organic matter in the substrate did not appreciably change from year to year, you might decide that this variable need only be monitored in five or ten year cycles.

Specific management questions that can be answered include:

- What level of function is achievable for natural wetlands and projects in a particular land use setting?

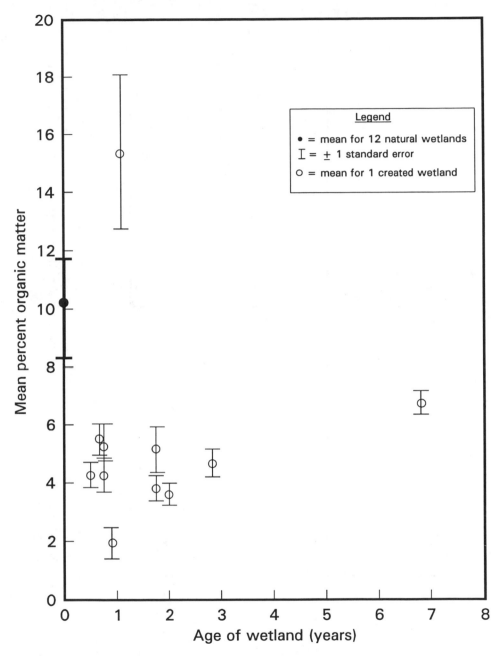

Figure 5-1. Performance curve generated using the mean percent organic matter in the upper 5 cm of soil from created and natural wetlands sampled in the Oregon Study. Mean for the natural wetlands is the grand mean from 10 soil pits sampled at each of the 12 sites. Mean for the created wetlands is for 10 soil pits sampled at each of the 11 sites. Organic matter was measured as ash free dry weight.

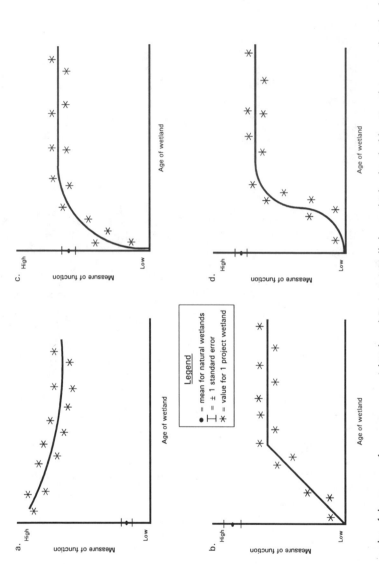

Figure 5-2. Examples of shapes a performance curve might take. a) Projects initially have a higher level of function than natural wetlands, but with time the level of function decreases and approaches the average for the natural wetlands; b) The level of function of the projects increases at a constant rate over time (linear relationship) and then levels off; c) The level of function of the projects increases quickly with time and then levels off (quadratic relationship); and d) The level of function of the projects increases slowly, then quickly, then levels off (logistic relationship).

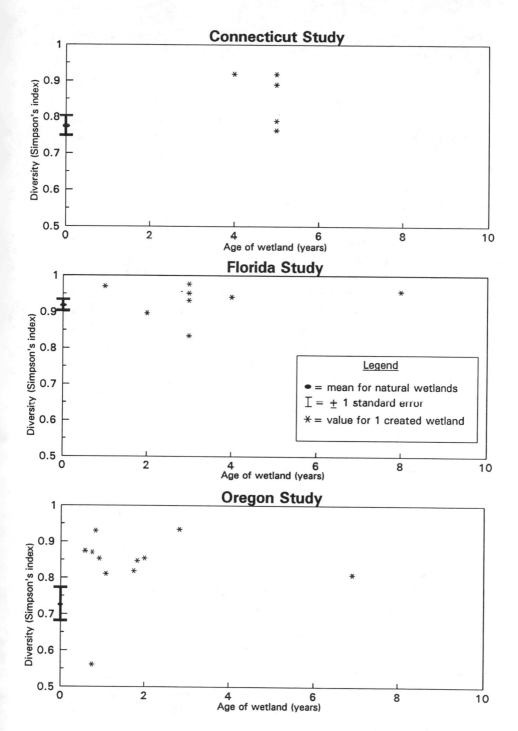

Figure 5-3. Performance curves generated using plant diversity data from the Connecticut, Florida, and Oregon Studies.

- Do the projects achieve the same level of function as natural wetlands?

- How long does it take for projects to achieve the desired level of function?

- How can monitoring be timed so as to obtain the most reliable information?

The WRP Approach is designed to evaluate wetland replacement in terms of the wetland function actually replaced and the time required to reach the level of function desired or possible. The information can be used a number of ways. For example, resource managers must include an evaluation of the potential performance of a project when deciding whether replacement or restoration is desirable. To accomplish this they need to distinguish between projects that replace the function of natural wetlands and those that do not (Figure 5-4, case D versus A, B, and C). They also need to know if a proposed improvement to a site has potential for improving the current status of the resource (Figure 5-4, case B). Finally, they must distinguish those projects that replace the function of natural wetlands in an acceptable time frame from those that take prolonged time to mature (Figure 5-4, case A versus case C). The time that it takes to replace functions is often ignored. Without the incorporation of such knowledge into management schemes, we risk losing the organisms and processes associated with mature, natural wetlands because critical features of the mature systems can be lost before sufficient time has passed for restoration and creation efforts to replace them.

Summary or Descriptive Graphs

Summary or descriptive graphs can be used to describe samples and identify outliers. There are many different types, but we found bar charts and box and whisker plots especially useful. In our studies, bar charts were used to compare a measure or indicator of function for a sample of created and natural wetlands. For example, Figure 5-5a shows the percent of open water for each of the created and natural wetlands sampled in the Oregon Study. A graph similar to a bar chart was used to compare the different weighted average scores found in the Florida Study by ranking them in order along a line of possible weighted average scores (Figure 5-5b). Box and whisker plots (Figure 5-5c) can also be used to describe a sample. The box outlines where the 25th (lower), 50th (median), and 75th (upper) quartiles of the data are located. The plot also gives an indication of the variability, symmetry or skewness of the data, and the presence of outliers.

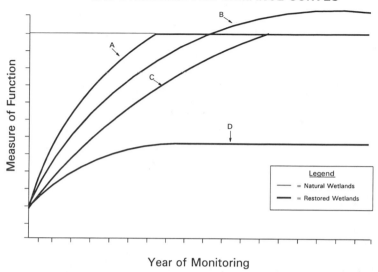

HYPOTHETICAL PERFORMANCE CURVES

Year of Monitoring

Figure 5-4. Hypothetical performance curves illustrating four different patterns of project development that could be used in making management decisions. A population of natural wetlands is being compared to four populations of projects (in this case restored wetlands) relative to a measure of wetland function. Population A develops more rapidly than populations B,C, and D. A and C achieve the same level of function as the natural wetlands, while population B exceeds the level of the natural wetlands and population D never achieves the level of the natural wetlands.

Time Series Graphs

Time series graphs are similar to performance curves. Both display levels of function versus time, but in the case of time series graphs, data points are not values for individual wetlands of different ages, but are observations from one (or more) wetland(s) sampled over time (Figure 5-6). These types of graphs work well with paired data (e.g., projects and natural wetlands in the same watershed). The similarity of projects and natural wetlands, and whether or not levels of wetland function change with time, are two pieces of information that can be ascertained from these graphs. In addition, if water levels are plotted versus time, the period when the water level is at, or above, the surface can be used to determine what portion of the site is jurisdictional wetland.

Characterization Curves

Characterization curves are also referred to as frequency distribution curves or histograms (Figure 5-7). They are a type of bar graph with the vertical axis representing frequency and the horizontal axis representing the variable of interest, usually grouped into classes. For example, we might plot number of wetlands versus an indicator or function of interest such as percent organic matter found in soil. If you use number of wetlands for the y axis, then

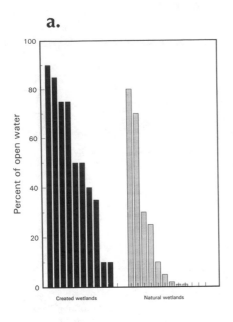

a.

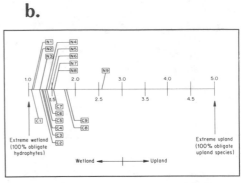

b.

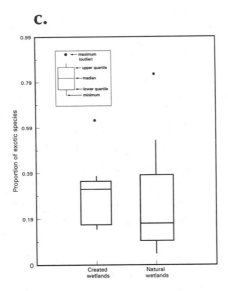

c.

Figure 5-5. Examples of summary or descriptive graphs. a) Bar graph of the percent of the site that was open water on 11 created and 12 natural wetlands from the Oregon Study. ONE BAR = ONE WETLAND b) Weighted average scores (Wentworth et al. 1988) for the type of vegetation found on individual created (C) and natural (N) wetlands from the Florida Study (adapted from Brown 1991). c) Box and whisker plot of the proportion of the plant community of created and natural wetlands from the Oregon Study that was composed of exotic species.

An Approach to Improving Decision Making in Wetland Restoration and Creation

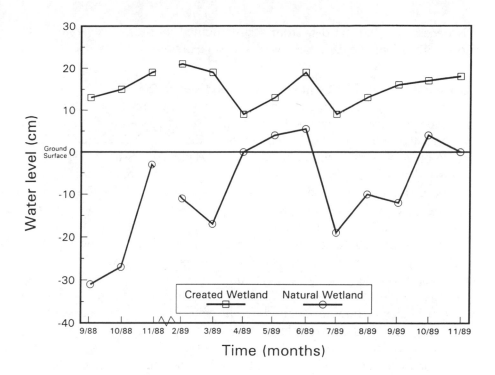

Figure 5-6. Monthly water levels (cm) for a pair of the created and natural wetlands from the Connecticut Study (adapted from Confer 1990).

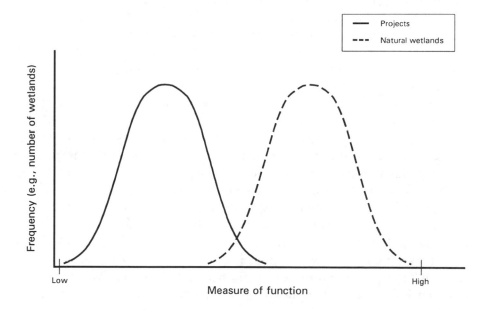

Figure 5-7. Example of hypothetical characterization curves illustrating how natural wetlands and projects might be compared.

it is necessary to collect data from a large number of sites to generate these curves. Therefore, they are not always a feasible option.

TECHNIQUES FOR DETERMINING DIFFERENCES IN SAMPLES

The four graphic techniques are useful for displaying the monitoring data so that patterns can be discerned. However, managers often want to take the data one step further and quantify differences observed when comparing projects to other projects or to natural wetlands. A number of statistical tests are available that can be helpful. It is beyond the scope of this document to give a detailed description of statistical tests and the assumptions for their use; however, we do provide brief descriptions of some tests and references for more detailed information. Many introductory texts give detailed explanations of the tests and procedures described in this chapter. Snedecor and Cochran (1980) is a good reference for statistical techniques such as comparing means, variances, and slopes, and calculating confidence intervals. Devore and Peck (1986) is another good reference for basic statistical methods that gives detailed examples of how to create different types of descriptive or summary graphs (e.g., box and whisker plots). Neter et al. (1990) is a good source for information on regression techniques, and Krebs (1989), Sokal and Rohlf (1981), and Ludwig and Reynolds (1988) give some specific methods for use with ecological data. This list is by no means exhaustive, but provides references for the tests if you need more information. In addition, statistical software that will perform the statistical analyses is available.

The data used for the statistical tests must fulfill various assumptions (e.g., normal distribution, independent observations, equal variances) for the use of the tests and the results to be valid. However, there are ways to manipulate the data (e.g., transformations) so that they meet the assumptions, and there are also alternative tests that can be used for different situations. Checking to see if the assumptions of a given statistical procedure are met is a necessary first step in data analysis and is described in the suggested references.

All four techniques—performance curves, summary or descriptive graphs, time series graphs, and characterization curves—can be used to compare samples of wetland projects with samples of other projects or natural wetlands, while statistical tests can be used to determine whether there are significant differences between samples. For example, you might be interested in whether the percent of organic matter in the substrate is different for a group of projects of the same age that were sampled in 1980, and another group of projects of the same age that were sampled in 1990. Tests of hypothesis including a Student's t-test or a nonparametric equivalent (e.g., Wilcoxon-Mann-Whitney rank-sum test) can be used to determine whether the means of the two groups are different.

Another statistical tool that you could use is a confidence interval. Hypothesis tests determine whether there is a difference between two samples while confidence intervals provide a range of values for the difference in means. For example, calculating a confidence interval based on projects sampled in 1980 and separate projects sampled in 1990 would give you a range of likely values for the difference in the population means between the two years. This could be important to know. The level of organic matter in the substrate is related to certain wetland functions such as water quality improvement. If the design of the projects sampled in 1990 had been modified from the design of the projects sampled in 1980 in an effort to accelerate the accumulation of soil organic matter, it would be helpful to quantify the difference between the two samples. Higher levels of percent organic matter in the sample from 1990 could indicate that the changes in the design did accomplish the goal. To make sure this was an actual improvement, other aspects of the projects should be examined to confirm that important wetland functions were not affected.

You could also compare the variability of two groups. Looking at summary graphs can give an indication of whether the variability of two samples is different. The range of the data can be determined from most graphs and gives an indication of the variability. Box and whisker plots can also be used for indicating variability. For example, in Figure 5-5c the box and whiskers associated with the natural wetlands are longer than the box and whiskers associated with the created wetlands. This indicates that there is more variability in the data from the natural wetlands. You could also use a statistical test to determine if there was a difference. If certain assumptions were met, tests such as an F-test or a Levine's test could ascertain whether the variability of two samples was the same. However, in most cases looking at summary graphs is sufficient, especially if the sample sizes are approximately equal.

When would it be relevant to determine whether the variability between two samples was similar or different? One situation would be to determine whether the sites within the samples were homogeneous. Another more specific example would be if you were monitoring water levels over time at a pair of project and natural wetlands. Larger variability in water levels found at either the project or natural wetland would be of interest due to the potential implications for how the site could perform a specific function such as wildlife habitat.

Multiple regression could be used to determine if there was a statistical difference between the steepness of two lines. This would enable you to determine whether the rate of change (or slope) on a performance curve from a sample of projects was similar to the rate of change on a performance curve from another sample of projects (e.g., a sample from more recent projects). Certainly, it would be reassuring if the slope of the performance curve from the sample of more recent projects approached the mean level of the natural wet-lands faster than the slope from the earlier sample (Figure 5-4).

Statistical differences do not always imply meaningful biological differences. You will need to decide how much of a difference between two groups is meaningful for your management or research needs. This points to the need for confidence intervals in addition to tests of hypothesis. However, you may simply want to rank the data to determine how specific wetlands compare with each other, or use other purely descriptive techniques for comparing samples. Ultimately your professional judgment as a manager and an ecologist is needed to interpret the tests and to decide if the results make sense. Results that are difficult to interpret or do not make sense could indicate problems with data collection, entry or analysis.

EVALUATING PROJECTS AND SETTING PERFORMANCE CRITERIA

The following example from the Oregon Study shows the process for setting performance criteria based on vegetation cover data that was collected in the field and then summarized in graphic displays. The wetlands studied were all located in the Portland Metropolitan Area and consisted of 12 natural and 11 created ponds with a fringe of emergent vegetation (see Figure 3-5).

The percent cover of each plant by species was estimated in 0.1 or 1.0-m^2 quadrats. The 0.1-m^2 quadrats were used in wetlands where the vegetation was short and relatively homogeneous. Forty quadrats were evenly spaced along transects that were placed to represent the different vegetation communities present. Percent cover was used to estimate standing crop.

The mean percent cover per quadrat was calculated for each created wetland and then plotted versus the age of the wetland (Figure 5-8). Since the age of the natural wetlands was not known, the mean cover per quadrat and its standard error were calculated for the sample of natural wetlands. Note that most of the created wetlands have values lower than the mean for the natural wetlands. We could use this information to set performance criteria. A possible criterion based on our data would be that the mean cover of emergent vegetation on created wetlands during the first three years of development would be less than that of similar natural wetlands.

We also tested this more formally with a Student's t-test using the mean values from the individual wetlands as observations. The question of interest was whether the mean cover per wetland was the same for the created and natural wetlands. The results of the Student's t-test indicated that the mean cover per wetland was larger for the natural wetlands (p=0.002). This result makes sense because the created wetlands were younger than the natural wetlands. However, we would expect that with time the mean cover on the created wetlands would increase to the level found on the natural wetlands. By re-

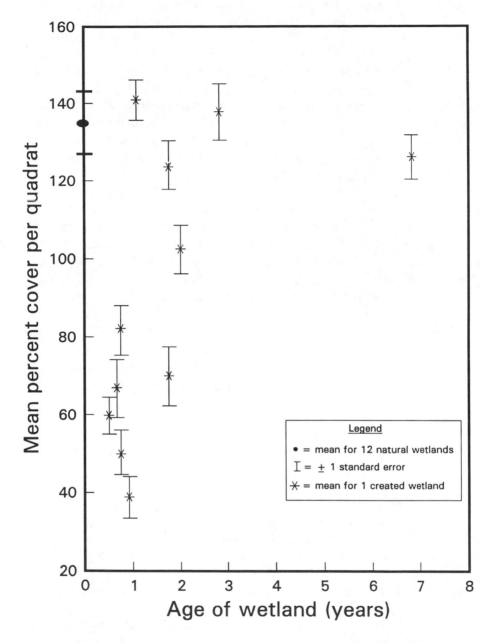

Figure 5-8. Mean percent cover for created and natural wetlands from the Oregon Study plotted versus project age. This is the beginning of a performance curve for this set of projects. Mean for the natural wetlands is the grand mean from 40 quadrats sampled at each of the 12 sites. Mean for the created wetlands is for 40 quadrats sampled at each of the 11 sites.

sampling the created wetlands in the future, we can get an indication of whether the increase is occurring and at what rate.

The cover data also are displayed using a box and whisker plot. By examining Figure 5-9, you can see that the variability of the mean cover of the created wetlands is larger than that of the natural wetlands. This would indicate that the sample of natural wetlands was more homogeneous in terms of cover than the sample of created wetlands.

The diversity of the species found on the created and natural wetlands also was calculated from the cover data. A performance curve of the data is shown in Figure 5-10. In contrast to percent cover, the diversity of species on the created wetlands tends to be greater than the mean diversity of species on the natural wetlands. Again, we would expect higher diversity because the created wetlands are relatively new areas, and it is typical for a variety of species to invade. We also would expect the diversity to decrease with time as cover and competition for resources increased. The performance criterion set from this data would be that the diversity of herbaceous vegetation on created wetlands during the first three years of development would be greater than or equal to that on similar natural wetlands. Because the two samples were not normally distributed, we tested the difference between the diversity of the created and natural wetlands with a nonparametric test (Wilcoxon-Mann-Whitney) and found that there was some evidence that they were different ($p=0.09$).

The presence of outliers can provide additional information. In Figure 5-10, there is one created wetland with lower diversity than the others. It is important to determine the reason for the low diversity and whether additional investigation is needed. For example, another site visit could be necessary, or rechecking the data sheets or field notes could be warranted. In the case of our example, as a first step we checked our field notes. They documented that this site was receiving petroleum run-off from the parking lot of a city operations plant. We hypothesized that the run-off was affecting plant diversity (as well as other attributes of the wetland) and that a site visit was needed to ascertain what should be done to correct the problem.

A created wetland also could be an outlier and indicate a potentially positive situation, if it had an unusually low value of an undesirable function or a high value of a desirable function. In these cases, we suggest that the wetland be examined more closely to see if there are any indications of ways to promote the development of the desirable characteristic. For example, suppose the only mulched project (see Chapter 6 for definition) in a group of projects of the same age, had a significantly higher percent organic matter in the substrate than did other projects in the group. Then, a resource manager might want to mulch future projects to determine if the mulching continued to enhance the organic matter in the substrate.

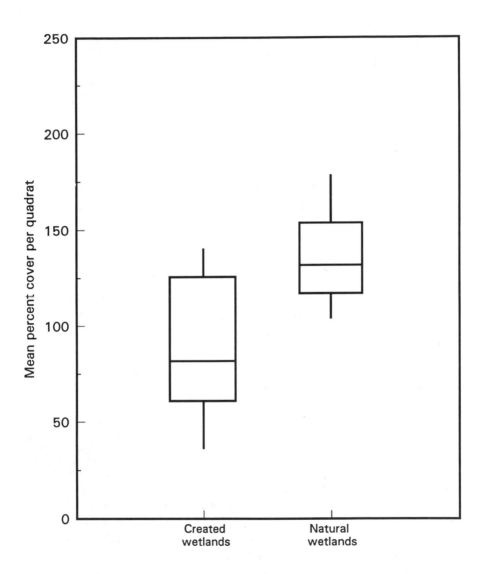

Figure 5-9. Box and whisker plot of cover data for created and natural wetlands from the Oregon Study.

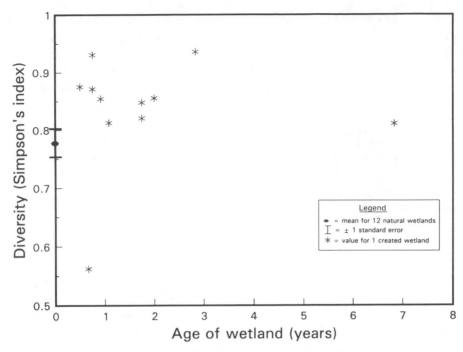

Figure 5-10. Performance curve of plant diversity data from the Oregon Study.

The vegetation data also were used to compare the percent of species that were common between the created and natural wetlands. We found that 41% of all the species were found on both the created and natural sites. Forty one percent of the species were unique to the created sites and 18% were unique to the natural sites. However, the percent of species in common between the created and natural wetlands was higher on a site-by-site basis (Figure 5-11). The mean percent of species in common for created and natural wetlands was 68.1% (s.e.=3.1). A 95% confidence interval for the mean would be approximately 62-74%. This interval could be used to evaluate the means from future samples of created and natural wetlands. As another option, we could determine whether future created wetlands fell approximately within the range of values of our sample (45-81%). We would be pleased if the created wetlands fell into the upper end of the range, or above it, because this would indicate that there were a greater number of species in common between the created and natural wetlands. The performance criterion set from this data would be that 62-74% of the emergent or herbaceous species found on a created wetland during the first three years of development should also be found on similar natural wetlands. These types of analyses, and the associated criteria, also could be used for cover data. In this case, the criterion might be—expect the average overlap of species' cover to be between a and b for the created wet-

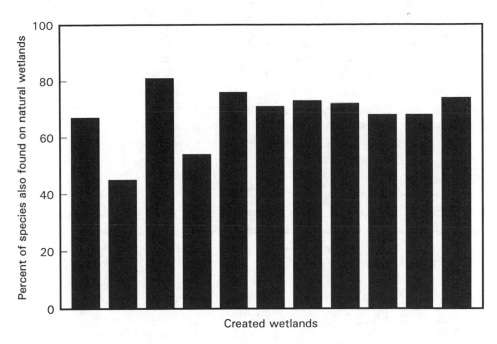

Figure 5-11. Bar graph of the percent of species overlap between individual created and natural wetlands from the Oregon Study (adapted from Gwin and Kentula 1990). ONE BAR = ONE CREATED WETLAND

lands during the first three years of development when compared to natural wetlands.

Finally, the vegetation data were used to compare the portion of species that were wetland plants. A weighted average, calculated using the wetland indicator status of the plant (Reed 1988, appropriate volume PNW) and its cover value, was determined for each wetland (Wentworth et al. 1988) (Figure 5-12). Note that the created and natural wetlands are interspersed along the wetland/upland axis and generally fall between the values of 1.3 and 2.8. This range indicates that there is a good to high probability that the sites are wetlands. However, additional data regarding soils and hydrology are necessary to determine conclusively that some of the sites are wetlands. A performance criterion set using this graph could be to expect values between approximately 1.0 and 3.0 for weighted averages calculated using data on herbaceous vegetation for both created and similar natural wetlands in the Portland Metropolitan Area. Because the weighted average value can be used to help determine whether the vegetation parameter for delineation is met, it could be useful to pick a specific cut-off value above which additional soils and/or hydrology data could be necessary for created wetlands.

As you can see from the discussion above, the vegetation cover data from the Oregon Study were used in many ways. We were able to display the data

---

Chapter 5: Evaluating the Data and Developing Performance Criteria

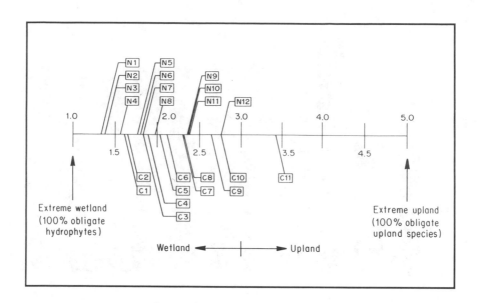

Figure 5-12. Weighted average scores (Wentworth et al. 1988) for the type of vegetation found on individual created (C) and natural (N) wetlands from the Oregon Study.

in different graphs and to examine the variables found within the data set (e.g., percent cover, diversity, percent wetland species). We used the information from the curves to answer questions such as: How does the species composition on the created sites compare with that on the natural wetlands?, and, Is the percent cover of vegetation similar for the created and natural wetlands? Finally, we used the information from the curves to set performance criteria that will aid in evaluating future created wetlands of the same type in an area.

The criteria that we assign will be useful to have in the future when we return to the Portland area to sample recently created wetlands. At that time, we will be able to compare the data from the recently created wetlands with the data from the original set of 11 created wetlands to determine if it takes less time for the more recent group to become like natural wetlands. We will also be able to determine if the recently created wetlands tend to fall within the criteria we set based on the original sample. If they do not, we can hypothesize as to why not. We will check for any new outliers, and use them to identify what to do, or not do. For example, we can determine whether the changes we made based on the outliers from the first group of created wetlands (e.g., the addition of mulch) were helpful. Sampling additional created wetlands will enable us to develop more precise performance criteria and begin to make decisions on whether the performance of the projects is adequately replacing ecological function.

At the same time that we sample the recently created wetlands, we plan to resample the original 11 created wetlands. This will give us additional data we can use to begin constructing the hypothetical performance curve illustrated in Figure 3-1a, which in turn will help us to set performance criteria for future projects.

An Extension of the Example

A resource manager might have different states or regions to compare. To illustrate how you could make such a comparison, the following example uses data from studies that compared created and natural ponds with a fringe of emergent marsh in Oregon, Connecticut (Confer 1990) and Florida (Brown 1991) (see Figures 3-5, 5-13, 5-14, respectively, for examples). Although, the Connecticut Study used paired wetlands, it is included in this example for illustrative purposes. Figure 5-3 shows performance curves of the plant species diversity found on the wetlands sampled in the three states. Note that for each of the studies, although the mean levels of diversity for the natural wetlands are different, the level of diversity for the created wetlands tends to be higher. The performance criterion developed using data from these curves would be to expect the level of plant diversity on projects during the first three to five years to be greater than or equal to the mean found on similar natural wetlands. A resource manager whose data looked like Figure 5-3 would recognize any project with a species diversity less than that of natural wetlands as a probable outlier requiring further evaluation.

We do not want to imply that the greater diversity found on the created wetlands is necessarily a desirable or lasting phenomenon. For example, the diversity could have been due to weedy, opportunistic, or exotic species, which points to the importance of evaluating the species composition. However, we consistently found the diversity to be greater in the newly created wetlands. Therefore, we would expect to find higher diversity in newly created wetlands of the same type and age in different parts of the country.

Example of How to Use Time Series Graphs

Graphs of surface water levels versus time, or of other variables where seasonal changes might be expected, can show trends over time and be used with paired wetlands or wetlands that are grouped in some meaningful way (e.g., wetlands that are hydrologically similar). Figure 5-6 shows a hydrograph from the Connecticut Study (Confer 1990). As you can see, the highs and lows in water level occur at approximately the same time in the two wetlands. However, the created wetland appears to have water levels that are less variable than the natural wetland. It also has a higher average water level than the nat-

Figure 5-13. Example of an emergent marsh in the Connecticut Study.

Figure 5-14. Example of a pond with a fringe of emergent vegetation from the Florida Study.

An Approach to Improving Decision Making in Wetland Restoration and Creation

ural wetland. How these two factors affect the fauna, vegetation, and other parameters associated with the wetland would be topics to pursue. These graphs can also help in establishing how much of the site is wetland. How many days the water table is within a set distance from the surface and/or how many days the water levels are at, or above, the surface are important factors in determining whether a site is a wetland.

Example of How to Use Characterization Curves

Figure 5-15 shows a histogram of the percent of organic matter found in the upper 5 cm of soil in both created and natural wetlands sampled in the Oregon and Florida Studies. We combined the data for this example because there were not enough data from each study separately to generate the curve. Since the data came from two studies, they are used for illustrative purposes only. You could use a display of this sort to determine the shape of the distribution of the created and natural wetlands, to compare the levels of function for the created and natural wetlands, and to document the amount of overlap between created and natural wetlands. After examining Figure 5-15, you might expect to see future created wetlands with soil organic matter values less than or equal to that of natural wetlands. This could be tested directly with a hypothesis test.

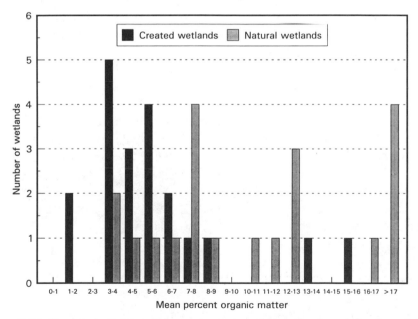

Figure 5-15. Characterization curve of percent organic matter measured as ash free dry weight in the upper 5 cm of soil. Data are from the Oregon and Florida Studies. Error bars have not been included since the data came from two different studies. The example is included for illustrative purposes only. Hypothetical curve is displayed in Figure 5-7.

SUMMARY

In this chapter we have presented specific ways to graphically represent data collected from monitoring wetlands. Four different graphs are suggested, each of which answers specific questions about projects and natural wetlands. The graphs are easy to create, and data can be added to them as they are collected. The more data collected and compiled, the more precise statements based on the information will be.

In this chapter we also have suggested the use of graphic representations of the data and statistical tests to develop performance criteria. The criteria can be used to evaluate the performance of projects based on the results of past monitoring. Criteria for different regions can be compared and general trends in the development and performance of wetland projects identified.

Finally, suggestions for improving wetland management can be made based on this information. Specifically, establishing performance criteria will aid managers in decision making when defining objectives, anticipating future problems, and planning for the long-term success of wetland projects. It also will enable managers to make knowledgable decisions as to when wetland restoration, creation, or enhancement are viable options.

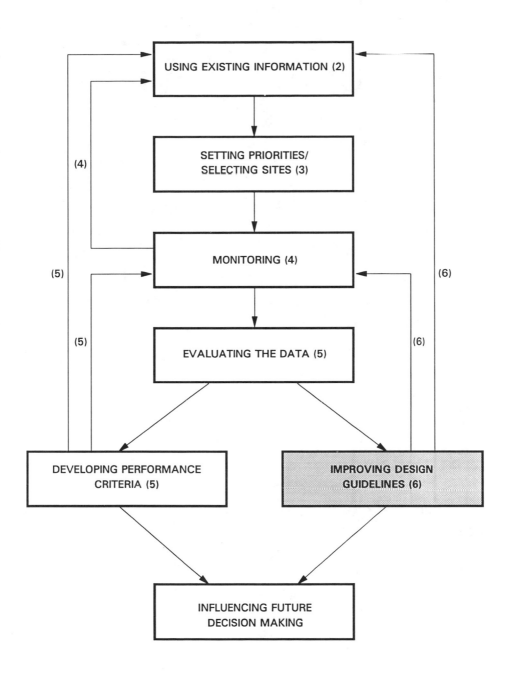

CHAPTER 6

Improving Design Guidelines

Information from natural and previously created or restored wetlands can be used to evaluate the design of current projects and improve the design of future projects. We present the results of several years of research in different areas of the country to illustrate how to identify when the design of wetland projects is or is not producing the intended results. This information can then be used to plan and design projects that have an increased probability of performing like natural wetlands.

We use two easily recognized wetland characteristics to present our case, wetland type and vegetation. Other characteristics of wetlands will be addressed, however not in great detail, because less information was collected in the field on these characteristics, and the corresponding design information from the project files was limited to nonexistent.

WETLAND TYPE

A significant finding of research conducted in Oregon (Kentula et al. 1992) and Wisconsin (Owen 1990) was that, although created wetlands tended to be located in the same county, river basin, or body of water as the associated impacted wetlands, there were differences between the wetland types impacted and those created. Therefore, local gains and losses of certain wetland types occurred. A similar trend has been observed on a national scale. A recent report by the FWS (Dahl and Johnson 1991) states that, although gains in some wetland types appear to offset some of the overall wetland losses that occurred from the mid-1970s to the mid-1980s, many gains were simply conversions between wetland types. Most significantly, gains occurred in non-vegetated unconsolidated bottom wetlands (i.e., ponds).

Determine if the Project is Typical of Wetlands in the Region

To use Oregon as an example, an analysis of the Section 404 permit record indicated that 23% of the wetlands created were ponds, but no natural ponds were impacted. In addition, an examination of the NWI maps for the Willamette Valley, Oregon, where most of the ponds were constructed, revealed that ponds were not a wetland type typical of the region. The only natural ponds found were associated with major water courses, and, therefore, were subject to yearly flooding. Typically, the created ponds were isolated hydrologically from rivers (Kentula et al. 1992).

Field research supported the analysis of the permit record, in that most wetland projects sampled in the Portland Metropolitan Area were ponds. The structural characteristics that defined the projects as ponds included steep banks sloping down to an expanse of open water with a fringe of wetland vegetation at the water's edge. Because the structural characteristics that define wetland type also influence wetland function, wetlands of different types are likely to perform different functions. Therefore, when making decisions about which wetland type to create or restore, one of the first considerations is, what are the most important functions to replace. If natural wetlands in the local landscape perform these functions, then the decision should be made to create or restore wetlands of the same types. If there are compelling reasons for creating a type different from that which occurs naturally (i.e., need for flood detention, sediment retention, or wildfowl habitat), the decision should be well thought out and potential consequences anticipated. The decision should not be based on what is the most convenient wetland type to create because of available land or financial limitations. Furthermore, because we do not know the ecological ramifications of replacing impacted wetlands with wetlands of different types (Kentula et al. 1992), reason suggests that we err on the side of caution and do our best to create types that occur naturally in the area. Undoubtedly, there are good reasons, geologically or hydrologically, as to why the natural wetland types occur where they do.

Influence of Bank Slopes on Wetland Type

Data collected at created wetlands in Oregon (Gwin and Kentula 1990) and Connecticut (Confer 1990) indicate that a large proportion were built with steep slopes and consequently, only narrow fringes of hydrophytic vegetation at the water's edge have become established. These created wetlands had notably greater areas of open water than did similar natural wetlands sampled in Oregon and Connecticut. The large area of open water and steep bank slopes of these projects resulted in ponds, rather than the palustrine emergent marshes that were planned. Ten of 15 mitigation projects studied in Wisconsin also resulted in ponds (Owen 1990).

Slopes and water depth influence the type and extent of wetland that will result from wetland creation efforts (Owen 1990). For example, the steepness of bank slopes leading into the wetland from surrounding areas influences the extent of the vegetation community. Steep slopes provide less area at the appropriate elevations and with appropriate hydrology for wetland vegetation to become established. A narrow fringe of wetland vegetation is likely to occur around a steep-sided pond, whereas, on a gentle slope wetland vegetation will occupy a broad expanse.

Data collected to compare created with natural wetlands in Florida indicated that, although the created wetlands did not have greater proportions of open water, they did have steeper bank slopes, greater basin depth, and consequently, greater mean and maximum water depths than similar natural wetlands in the region (Brown 1991). The differences in slope between the natural and created wetlands were probably the result of a combination of inadequate design and economic realities associated with the high value of real estate. The created wetlands were built in residential or commercial developments (this is also true of projects sampled in the Oregon Study), often tucked into corners, beside roadways or associated with storm water systems. The lack of space and possible unwillingness of the landowner to commit larger land areas (because of the high development value of the land) may have contributed to the pattern observed where the amount of land needed for the wetland was decreased by increasing the slope of the banks. Gentle slopes require a greater amount of land to achieve adequate basin depths and result in larger transitional area surrounding each wetland (Brown 1991).

It is unlikely that ponds can replace the lost functions and values of wetlands that are filled (Owen 1990), and, as stated above, they often represent a wetland type that does not exist naturally in the area. However, ponds are a simple and inexpensive type of wetland to construct (McVoy 1988, Novitzki 1989), and are often favored over other types of wetlands because of potential waterfowl habitat values (Gene Herb, Oregon Department of Fish and Wildlife, Forest Grove, Oregon, personal communication). Also, because ponds can be tucked into small places, they are often the wetland type of choice when there are constraints due to the amount or value of land available for a project. When there is only a small piece of property available for the project, it is often decided that a wetland of the same size as the property will be constructed. Unless the ground surface is at or very near the water table (in this case, the area may already be a wetland!), steep bank slopes will be required to construct the desired area of wetland. Usually, this procedure will result in the creation of ponds.

A literature search found that most experts recommend that bank slopes for created and restored wetlands range between 5:1 to 15:1 horizontal to vertical (H:V). However, recent research (Brown 1991, Gwin and Kentula 1990, Owen 1990, Confer 1990) has indicated that bank slopes for most wetland types should be constructed at or beyond the gentle end of this range (somewhere near 15:1 H:V or flatter) to make the wetland projects more similar to natural wetlands. The slopes of ten of the twelve (83%) natural wetlands measured in Oregon were flatter than 10:1, and the slopes of six of these ten were flatter than 20:1. Gentle slopes, that occupy a large expanse of the area between the upland and any inundated area, allow development of a wide expanse of wetland vegetation rather than a narrow "ring around the pond" of vegetation at the water's edge. Figure 6-1 illustrates the difference in bank slopes and Figure 6-2 illustrates the difference in the topographical profiles between created and natural wetlands in the Oregon Study.

The slopes of natural wetlands can be used as guides for contouring wetland projects. Therefore, to design a project to have topography similar to that of natural wetlands in the region, we suggest the following steps.

- Select a random sample of natural wetlands similar to the project type;

- At each natural wetland in the sample, establish transects to determine the topography. A method for determining transect placement in wetland areas along watercourses is presented by the Federal ICWD (1989). The objective is to place the transects so that they are representative of the site. Whatever method is used for determining transect placement, the decision process should be documented;

- Measure and record relative elevations at predetermined intervals along each transect with transit and stadia rod. The measurement interval along each transect will depend upon the size of the wetland, the steepness of slopes, and microtopography; and

- For each natural wetland sampled, determine the "zero" point from the lowest elevation measured, and adjust all elevations relative to zero.

Relationship Between Bank Slopes and Wetland Area

Once the bank slopes for the desired wetland type have been determined, the next step is to decide the amount of area required to construct the wetland. In the following sections we will describe: 1) the factors to consider when determining how much land will be required for the wetland project; 2) the design of the basin when a sufficient amount of land is available; and 3) the design of the basin when the amount of land is limited.

Figure 6-1. Pictures of typical natural (a), and created (b) wetlands in the Oregon Study.

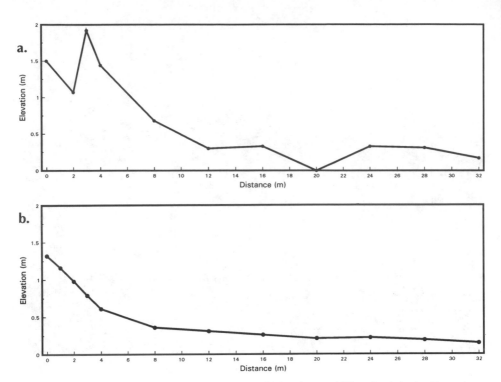

Figure 6-2. Topographical profiles for typical natural (a) and created (b) wetlands in the Oregon Study.

Determine how much land will be required

As discussed earlier, a 1-ha wetland will not fit into a 1-ha piece of land unless bank slopes are steep or water tables are at or near the soil surface. Therefore, if the project is intended to be a type other than a pond or steep sided basin, sufficient land must be set aside to include gentle bank slopes (which provide a transitional area between upland and wetland) and buffers. The required slopes, in conjunction with the vertical distance from the precon-struction ground level to the water table or water source, determine the amount of land required for the desired wetland area. The greater the distance to the water table, the larger the project site must be to reach the water table via gentle slopes. The pattern of the past has been to fit the project to the available site rather than to fit the site to the type of wetland desired. This phi-losophy must be changed—the project site must be chosen to accommodate the size and structural characteristics of the desired wetland type. Determina-tion of project size requirements by this method may be more legally defensi-ble than the ratios of wetland acreage to be created to offset wetland acreage destroyed that are currently used. Also, factors constraining the amount of land available for the project must be considered in some situations. If the project will be located within a highway meridian or residential or commercial development, there will be constraints on the amount of land available, and the type of wetland that can be created.

Design when adequate land is available

Information from our studies in Oregon, Connecticut, and Florida, as well as other current research, can be used to make better informed decisions about the type of wetland and the features of the basin design for your project. First, determine how much land is available for the project. If the available land is adequate to construct the desired wetland type with the appropriate basin characteristics, continue on with this section. If you will have only a small area, the next section describes the design of a basin when land is limited.

Decide what proportion of the available land will be inundated during the driest time of the year, what proportion will be vegetated with hydrophytes, the width of bank slopes needed to provide a transitional area from upland to wetland, and the width of buffer areas. Use the proportions found within natural wetlands in the area as your guide. Use the slopes of banks typical of local natural wetlands as a guide for contouring the basin of the project. For example, because the natural palustrine emergent wetlands sampled in Oregon typically had bank slopes flatter than 10:1, we would recommend contouring projects of this type in Oregon with a variety of slopes 10:1 and flatter.

The next piece of information you will need is the vertical distance from the water table or water source to the existing ground surface at the driest time of the year. This distance can be determined with water wells or by digging soil pits, and is the minimum depth the ground surface must be excavated to get water on the site if the project is to be supported by groundwater. Then, from your earlier decision on what proportion of the project should be inundated during the wet (and dry) times of the year, you can decide where the banks of the wetland should meet the water table.

You now have the information from which to determine how much land will be required for the project. For example, if 15:1 (H:V) slopes are required, the water table is 3-m below the ground surface, and you have decided that the banks should meet the water table at the bottom of the slope, a bank length of approximately 45-m will be required to inundate the wetland at the proper elevation via the required bank slopes. This distance must be added to all sides of the inundated area extending out toward the surrounding uplands and must be considered when determining the amount of land required (Figure 6-3).

Design when land area available is limited

If you must "fit" your project into a small area because, for example, it is ancillary to a residential or commercial development, or must be located near a roadway or within a highway meridian, you still have several options. The

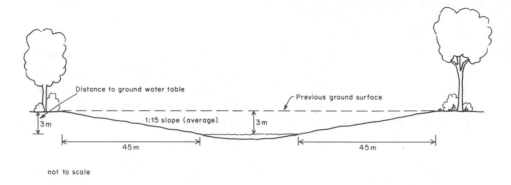

Distance to ground water table

Previous ground surface

3 m

1:15 slope (average)

3 m

45 m

45 m

not to scale

Figure 6-3. Illustration of how to determine the amount of land needed for creating a wetland given the bank slopes and the depth from the ground surface to the water table.

option most often used in the past was to construct a steep sided pond. However, we have the following recommendations that, if followed, should increase the proportion of the site that will be vegetated.

Bank slopes should be as gradual as possible. This will increase the amount of transitional area and the possibility for vegetation zonation along the moisture gradient extending down from the upland edge, despite seasonal and annual variability. Second, gradual slopes will help to stabilize the banks. If slopes are too steep, the banks may erode or fail and slump into the wetland, causing that area of the wetland to be at an elevation higher than planned, and consequently affecting the hydrology (Figure 6-4).

The wetland basin should be deep enough to attain the desired hydroperiod for the intended vegetation community (Hollands 1990), but not so deep that a pond will dominate the project. This means that the bottom of the wetland must be at an elevation where it will not be completely inundated, or where it will be inundated only very shallowly, so that emergent vegetation can persist.

VEGETATION

One of the significant findings of the comparison of created and natural wetlands in Oregon (Gwin and Kentula 1990) and Florida (Brown 1991) was that the composition of vegetation communities on the created wetlands was not notably different from the composition of the communities occurring on natural wetlands. Conspicuous differences did occur, however, between the composition of the vegetation communities on the created wetlands and the species included on the planting lists for those wetlands (Gwin and Kentula 1990, Gwin et al. 1991).

Figure 6-4. Erosion occurring on steep unvegetated banks at a created wetland sampled in the Oregon Study.

Example from the Oregon Study

Comparisons between the vegetation that occurred on created freshwater emergent wetlands in Oregon with the planting lists included in project plans found very few species in common (Gwin and Kentula 1990). The percentage of all species found on a project that were included on the corresponding planting list ranged from 0% to 7%. Therefore, between 93% to 100% of the species that occurred on each project were volunteers. This finding suggests that it may be unnecessary to plant freshwater emergent wetland projects. However, before this inference could be made, we needed to determine if the species that volunteered on the projects also occurred on natural wetlands, or if the vegetation communities of the wetland projects consisted mostly of inappropriate species such as invasive exotics.

Therefore, we compared the species that occurred on the created wetlands with the species that occurred on natural wetlands, and found that between 54% to 81% of the species were common to both groups. This suggests two things: 1) the species included on the planting lists were inappropriate for either the wetland types or the geographical area; and 2) planting lists should include the volunteer species because these species also occurred on natural wetlands in the area.

Example from the Florida Study

Comparisons similar to those in the Oregon Study were made between the created wetlands and their plans in the Florida Study (Gwin et al. 1991). The results were similar to those found in Oregon. Vegetation communities at three of the nine freshwater emergent wetland projects sampled were composed completely of volunteer species. For the remaining projects, between 85% and 90% of the species found were volunteers. The species that occurred on the projects were then compared with the species that occurred on natural wetlands of the same type and size in the same area. This comparison indicated that the percentage of species on the created wetlands that also occurred on natural wetlands ranged from 38% to 61% (Figure 6-5a).

The analysis of vegetation was taken one step further in the Florida Study. In addition to examining the species composition, we examined the relative abundance of each species. For those wetland projects that were planted, the percentage of the plant cover composed of species to be planted ranged from 0% to 33%. The majority of the plant cover, as well as the number of species on the project, was composed primarily of volunteer species. In addition, the percentage of the plant cover on the created wetlands composed of species that also occurred on the natural wetlands sampled ranged from 48% to 93% (Gwin et al. 1991) (Figure 6-5b).

Guidelines for Revegetation of Wetland Projects

With the patterns described above in mind, we developed a generic approach to the revegetation of freshwater emergent wetland projects. The primary objective is to vegetate the projects with species appropriate to the desired type of wetland (i.e., palustrine emergent marsh, riparian system, shrub/scrub wetland, etc.) in any given region. The species that occur in natural wetlands and that have volunteered on previously created or restored wetlands form the ecological "blueprint" for revegetation. In addition, permit conditions and specific project objectives are considered. To illustrate the approach using the results of the Oregon Study, we developed a partial planting list for palustrine emergent marshes of the Willamette Valley which is discussed in the following sections. Structural components of the project (e.g., hydrology, slopes, soils, etc.) are assumed to be correct.

To Plant or Not to Plant?

Planting can be very costly, and in some cases may be unnecessary. Therefore, we begin with an analysis of whether or not to enhance or accelerate revegetation by planting. The factors that contribute to the ability of a project to revegetate with appropriate wetland species include:

An Approach to Improving Decision Making in Wetland Restoration and Creation

a.

PERCENT OF SPECIES

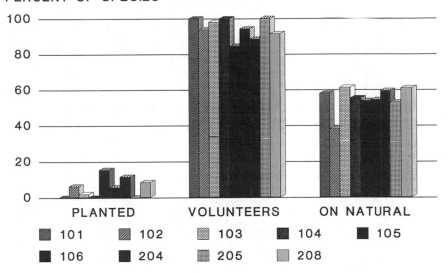

PLANTED   VOLUNTEERS   ON NATURAL

■ 101   ▨ 102   ▦ 103   ■ 104   ■ 105
■ 106   ■ 204   ▦ 205   ▨ 208

b.

PERCENT COVER

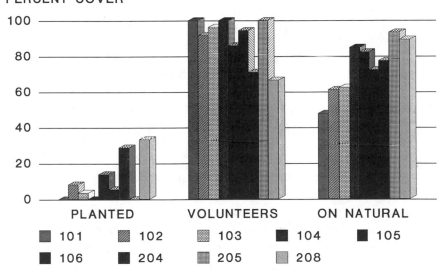

PLANTED   VOLUNTEERS   ON NATURAL

■ 101   ▨ 102   ▦ 103   ■ 104   ■ 105
■ 106   ■ 204   ▦ 205   ▨ 208

Figure 6-5. Comparison of the number (a) and the percent cover (b) of species found on created wetlands in the Florida Study with what was listed in the project plans and what occurred on similar natural wetlands.

- surrounding land uses and their contributions to the project in terms of pollutants and undesirable seeds (lawns, crops);

- isolation of the entire project, or a portion of it, from other wetlands and appropriate seed sources (e.g., the centermost portion of a very large project may require planting even if it is adjacent to another wetland and appropriate seed source);

- vegetation strata, specifically whether herbaceous or woody species are targeted to colonize the wetland (e.g., many herbaceous species volunteer and establish quite rapidly and, therefore, may not require planting; woody species often take longer to establish and, therefore, may require planting);

- time of year that construction takes place;

- hydrology, specifically timing and duration of inundation, water level fluctuations and flushing of the site; and

- soils present on the site or any soil augmentation (such as topsoil and plant propagules taken from a destroyed wetland).

If the project is located downstream, adjacent to, or nearby an existing vegetated wetland, it is highly likely the project will have the ability to revegetate itself. As described above, our research indicates (Gwin and Kentula 1990, Gwin et al. 1991) that even when newly constructed projects were planted, a high percentage of the species that occurred on the sites were volunteers, and a large percentage of these species were the same as those that occurred on local natural wetlands (Figure 6-5a). Therefore, although the time required for a project to revegetate without planting may be longer than with planting, if conditions are correct, it may be appropriate to allow the project to revegetate naturally.

If you decide not to plant, revegetation can be accelerated by mulching the project with soils salvaged from a destroyed "donor" wetland, known as salvaged marsh surface (SMS; Owen et al. 1989). Mulching with SMS can accelerate revegetation by providing seeds and other propagules, an organic surface horizon, and soil microflora (Kruczynski 1990). It also makes the substrate more conducive to rapid revegetation by reducing the evaporation of soil pore water, runoff, soil loss and erosion, and surface compaction and crusting (Thornburg 1977). Although propagules contained in the topsoil removed from the donor wetland should germinate on the project, a direct correlation cannot be drawn between the vegetation that was present on the donor wetland and the species that germinate from the seed bank contained in the SMS. Studies have shown that the species that germinate from the seed bank are often different from those that were present on the donor wetland (van der Valk and Davis 1976, Weinhold and van der Valk 1988). Species generally referred to as "mudflat annuals" most commonly germinate during the first few growing seasons after the wetland's constrution (Weinhold and van der Valk 1988). However, in subsequent years, other species contained within the seed bank should germinate and become part of the community.

Mulching has the potential to cause problems, however. Propagules from species different from those that occurred on the donor wetland, or from undesirable species may occur in the SMS. In addition, mulching may be unsuccessful if the propagules were damaged during the excavation and stockpiling processes.

If the project will be allowed to revegetate naturally, a monitoring program should be instituted within a year after construction to ensure that the project does revegetate with desirable species. The monitoring program need not consist of intensive sampling, but merely frequent routine checks to determine if the project is becoming vegetated and what the dominant species are (Chapter 4). If the project shows little sign of revegetating, if large areas of the site are being affected by erosion, if important components of the desired vegetation community are missing, or if many of the species are undesirable, a change in plan may be warranted and a planting scheme instituted.

Generating a Planting List

The vegetation community desired on projects should include those species and communities that occur on local, natural wetlands of the same type. In addition, the community planted on the project should be "low maintenance" , i.e., it should be composed of plants that grow well and reproduce at the given location in the particular climate with minimum care, and remain free of serious disease or insect pests (Stark 1972). The following sections describe how to: 1) use the composition of the communities on natural wetlands to generate a list of commonly occurring species; 2) determine which of the commonly occurring species are commercially available; and 3) generate a planting list from among the appropriate commercially available, commonly occurring species. Finally, we give the reader additional guidelines to help ensure the success of the planting strategy.

What species commonly occur on wetlands in the area?

Conduct a survey of the vegetation communities present on natural and previously created and restored wetlands of the target type within the local area. Identify all commonly occurring or dominant plants to genus and species. In general, dominant species are those that contribute more to the characteristics of the plant community than other species present, or that exert a controlling influence on, or define the character of, the vegetation community. Specifically, dominant species are those species in each vegetation stratum (i.e., trees, shrubs or herbaceous layer) that, when ranked in decreasing order of abundance and cumulatively totaled, exceed 50% of the total dominance measure (i.e., percent cover) for that stratum, plus any additional plant species comprising 20% or more of the total dominance measure for the stratum (Federal ICWD 1989).

From the vegetation survey, generate a list of the species that commonly occur on the natural wetlands and existing projects. This list reflects the species that may be appropriate to plant on local projects.

Which species are commercially available?

Check with local nurseries (especially those that specialize in native species) to determine which species are commercially available. Compare this list with the list of species that commonly occur on natural wetlands and existing projects to refine and narrow the list of species for planting. Then, determine the materials to be used for planting, (i.e., seeds, sprigs, culms, rhizomes, potted seedlings, or woody cuttings) based on commercial availability. Also consider the site conditions, how the plant materials are likely to be installed, and the goals of the project (e.g., aesthetics, wildfowl forage, etc.). Seeds have the least initial cost, but are likely to be lost from erosion or predation, and, therefore may be more costly in the long run. Sprigs, culms, and seedlings have a higher initial cost, but are more likely to establish successfully after transplanting (Garbisch 1986).

In addition to identifying which species are commercially available, it is important to ensure that the nurseries propagate their own plants and do not routinely remove them from natural wetlands in the area. However, it is also important that the nurseries acquired their original stock locally (within the study region). For example, until recently, many nurseries in the Pacific Northwest acquired wetland species from nurseries in the Midwest (Rex van Wormer, Independent Ecological Services, Olympia, Washington, personal communication). Although a certain species may occur locally, if the plants were acquired from a nursery in a region with a different climate, they may not survive.

Narrow the list of species to generate a planting list.

Use the appropriate Regional List of Species that Occur in Wetlands (Reed 1988) to determine which commonly occurring and commercially available species are:

- wetland species (obligate wetland, facultative wetland, facultative);

- endemic (i.e., native to the region); and

- exotics that are not invasive and are part of the natural community (i.e., those exotics that typically occur on natural wetlands but do not displace native species).

Use this information to generate a sublist that consists of only those plants that are wetland species, endemic, and/or noninvasive exotics. Then, use a regional flora to determine which species on this sublist are:

- herbaceous, shrubs, trees;

- weedy or opportunistic species;

- pioneering or early successional species;

- common, hardy, or rare species; or

- invasive (AVOID THESE).

The goals of the project will help determine which of these species should be chosen for planting. For example, although mostly herbaceous species should be chosen for planting a palustrine emergent marsh, some shrubs and/or trees might be chosen for planting along the tops of banks or to act as a visual buffer. If soil stabilization is a concern, it might be wise to choose a few species known for their rooting capabilities.

The above process will provide a list from which to choose species to plant on the project. See Table 6-1 for an example of the types of information this list might contain. The final list should include a minimum of plant species adaptable to the various elevation zones within the project, diversification will occur naturally. Garbisch (1986) recommends that one should:

- Select herbaceous species with potential value for fish and wildlife and with rapid substrate stabilization to help with initial establishment;

- Phase the establishment of woody species to follow that of the herbaceous species and determination or stabilization of water levels;

- Select species that are adaptable to a broad range of water depths. To determine this, decide from the vegetation survey which species come from wetter or dryer sites. In addition, information should be

Table 6-1. Partial list from which to choose species for planting on wetland projects in the Willamette Valley, Oregon. List was generated by following Steps 1-5 listed in the section Generating a Planting List. Wetland indicator codes were taken from the regional list of plant species that occur in wetlands (Reed 1988) and from consultation with LaRea Johnston, Assistant Curator of the Oregon State University Herbarium. Wetland indicator codes are: o—obligate wetland species; w—facultative wetland species; f—facultative species; p—upland species; i—introduced species; n—native species. The symbol + indicates the species is toward the high end of the category (more frequently found in wetlands); - indicates the species is toward the low end of the category (less frequently found in wetlands); and \ indicates the species is intermediate within the category. Notes and characteristics of the species and common names were derived from Hitchcock and Cronquist 1981, Steward et al. 1963, and Newling and Landin 1985).

SCIENTIFIC NAME	COMMON NAME	HABITAT CODE	NOTES/CHARACTERISTICS
Emergent Forbs:			
Alisma plantago-aquatica	Water plantain	o\n	Widespread in North America; shallow water & wet areas; good food source for wildlife.
Iris missouriensis	Western iris, Rocky Mountain iris	w+n	Pale to deep blue flowers; wet meadows and streambanks, but can tolerate dry summer condtions; B.C. to Calif & east to S. Dakota.
Submergent or Floating Forbs:			
Lemna minor	Duckweed	o\n	Temperate & subtropic freshwater lakes, ponds & slow moving water; good waterfowl forage.
Oenanthe sarmentosa	Water parsley	o\n	Widespread along Pacific Coast, Alaska to Calif.; still or sluggish water.
Forbs:			
Geum macrophyllum	Large-leaved avens	w+n	Widespread on wet ground from near sea-level to subalpine; Alaska to Baja Calif. east to Rocky Mtns.
Lysichitum americanum	Skunk cabbage	o\n	Mephitic; swamps & bogs from Alaska to Calif. & east to Idaho; easily propagated by division of underground stems.
Grasslike:			
Carex aperta	Columbia sedge	w\n	Wet lowlands, esp. floodplains; B.C. to NW Oregon & east to Idaho & NW Montana.
Carex obnupta	Slough sedge	o\n	Wet ground or standing water; Cascades to coast, B.C. to Calif; soil stabilizer & wildlife forage.
Ferns:			
Athyrium filix-femina	Lady fern	f\n	Very common, lowland to montane, circum-boreal; woods, meadows and swamps; can be a pest.
Shrubs & Trees:			
Corylus cornuta	Filbert, Hazelnut	p\n	Widespread at low elevation on well drained soil, B.C. to Calif & east to Idaho.
Fraxinus latifolia	Oregon Ash	w\n	Deep fertile moist soil, esp. streambanks; B.C., west Cascades to Sierran & coastal Calif.

An Approach to Improving Decision Making in Wetland Restoration and Creation

available from the nursery. Most nurseries will be happy to provide specifications for planting (e.g., in damp soils, depth of water, etc.);

- Avoid choosing only those species that are foraged by wildlife expected to use the site. Muskrat "eat-outs" of Typha occurred at two created wetlands sampled in the Connecticut Study (Confer 1990), resulting in a complete loss of vegetation at these sites. The vegetation at one site recovered and became more diverse. However, the vegetation at the other site did not recover immediately, and due in part to excessive flooding, the site remained unvegetated through the next growing season; and

- Avoid committing significant areas of the site to species that have questionable potential for successful establishment.

OTHER IMPORTANT STRUCTURAL CHARACTERISTICS

Although we recognize the importance of hydrology and substrates to the success of each wetland project, information from our research that can be applied to project design is limited. Therefore, our discussion of these features will not be as detailed as the previous discussion. We begin with examples of results from our research and follow with information from the literature on wetland restoration and creation.

Hydrology

Although the importance of hydrology to wetland restoration and creation is well recognized, it is often given only cursory attention in wetland design and construction. Research comparing created wetlands with their design and construction plans in Oregon (Gwin and Kentula 1990) indicated that most plans gave little consideration to the hydrology of the wetlands they were to create (Table 6-2). Also, field estimates indicated the proportion of each created wetland inundated by open water was considerably greater than that of natural wetlands in Oregon ($p = 0.03$) and Connecticut ($p < 0.0001$, Confer 1990). The created wetlands in Connecticut also had greater mean water depths than did natural wetlands, and had significantly less monthly fluctuation of water depths ($p < 0.05 - 0.01$, See Chapter 5). The reduced water fluctuations suggest that the hydrologic characteristics of the wetland projects differ from those of similar natural wetlands, and it is likely they will not develop or function as do the natural wetlands (Confer 1990).

To increase the likelihood of success, the hydrologic characteristics of the wetland project should mimic the hydrology (hydroperiod, water depths, amount of fluctuation, etc.) of local natural wetlands. This will require determining the hydrologic characteristics of the site chosen for the project, and de-

Table 6-2. The hydrology planned for created wetlands studied in the Portland, Oregon metropolitan area in 1987. Information was taken from the U.S. Army Corps of Engineers (COE) and the Oregon Division of State Lands permit files. EPA = U.S. Environmental Protection Agency.

SITE	HYDROLOGY INTENDED AT CREATED WETLAND
C1-CC	Letter from EPA to COE states that a hydraulic connection must be maintained between the project site and the adjacent creek to maintain adequate stream flow for fisheries.
C2-TI	Design plan shows a pipe leading into the created wetland from boat basin in the Columbia River.
O3-NS	Special Condition 8 of Attachment A to Permit states: ". . . connect newly dug wetland into the existing stream."
O4-MHP	Lake to receive water from two streams entering at its NW and SW corners. The streams drain a 572.1 acre watershed.
	Well water is to be supplied to the lake during seasonal low stream flow to maintain the water depth at agreed upon levels.
C5-MG	Drawings show a culvert leading into the wetland from under nearby street.
	Excavation to the level of an adjacent stream area subject to stream overflows and possible periods of standing water.
C6-3I	Existing creek channel to be rerouted through created wetlands. Stream flow estimated as about 4 cubic feet per second.
C7-SML	New stream channels to be excavated to increase stream length and supply water to project. Existing stream channels to be maintained as overflow channels.
C8-BSP	Drawing shows overflow slough connecting pond with nearby creek.
	The overflow channel is to be created between the existing overflow slough and the SE corner of the project site.
	"Roof water" will be discharged from two buildings into the pond.
	Surface waters from the surrounding developments to be discharged into the basin through diffuser pipes.
	Text states that "there may always be a slight freshwater flow from sub-surface seepage".

An Approach to Improving Decision Making in Wetland Restoration and Creation

signing the project to relate to the hydrology of the site. For example, the construction of hydrologically isolated ponds in areas such as Oregon where natural ponds are usually connected to a body of water (Kentula et al. 1992), will cause the ponds, although structurally similar to their natural counterparts, to function differently hydrologically. For the created ponds to function in a manner similar to that of natural ponds in this area, they must be hydrologically connected to streams or rivers. In addition, creation of a structurally similar wetland on substrate different from the substrate of natural wetlands may not facilitate similar hydrology because of differences in permeability (O'Brien 1986).

A substantial amount of hydrological information can be obtained from local natural wetlands with a modest investment in supplies and equipment. Water levels can be recorded continuously with water level recorders, or by reading a staff gauge during periodic site visits. With water level data, most hydrologic variables can be determined—hydroperiod, flooding frequency and duration, and water depths (Mitsch and Gosselink 1986). The development of water budgets for natural wetlands may further increase the probability of successfully creating or restoring a wetland, because the water budget provides a design for the hydrologic characteristics (Novitzki 1982). However, because a water budget is based on inflows equalling outflows, great care must be taken to ensure all components of the equation are accurately measured, and potential errors and their causes are estimated (Winter 1981). A typical water budget equation is: $P + OF + SWI + GWI = ET + SWO + R$, where

P   = precipitation on the wetland in inches or centimeters,

OF   = overland flow into the wetland,

SWI   = stream flow entering the wetland,

GWI   = groundwater inflow to the wetland,

ET   = evapotranspirative losses from the wetland,

SWO   = stream flow leaving the wetland, and

R   = recharge from the wetland to groundwater.

Other hydrological data to collect on local, natural wetlands include: flow conditions (i.e., whether water is flowing over the site or whether it is mostly stagnant, and whether it flows quickly or slowly); whether the flow of water is channelized or sheet flow; whether the ground was inundated or saturated at some distance from the surface or at the surface; the proportion of the wetland that is covered with open water; seasonal water level fluctuations; and locations and types of water inflows and outflows. In some regions certain of these data may already exist and be used in project design. For example,

Golet et al. (In press) document normal water level fluctuations in red maple swamps in Rhode Island, and Kantrud et al. (1989) describe the hydrologic regime of prairie basin wetlands in the Dakotas. It is very important that the hydrological characteristics of the project are documented in the design and construction plans for determinations of compliance and so that successful projects may be used as models for future hydrological design.

Soils/Substrates

Data collected in the Oregon Study showed that most created wetlands had significantly lower soil organic matter than did natural wetlands. The average percent organic matter in soils of projects (5.49% at 5-cm depth, S.E. = 1.05%) was significantly lower than that of soils of similar natural wetlands at depths of 5-cm (10.13%, S.E. = 1.67%), 15-cm, and 20-cm (p=0.002, p=0.02, p=0.02 respectively). Due to the young age of these projects, the lower organic matter was expected. What was unexpected was that one created wetland had organic matter much higher than all the other created wetlands and the mean for the natural wetlands. Further examination of this project could lead to insights into how to accelerate the accumulation of organic matter on other projects, concurrently increasing wetland functions related to soil organic matter content.

Soil organic matter is an important potential source of available nitrogen (Langis et al. 1991). In addition, soil organic matter stores nutrients and provides organic substrates for bacteria involved in nitrogen fixation, denitrification, and the sulfur cycle (PERL 1990). The lower soil organic matter of the created wetlands suggests that these soils have less energy for soil microbes to recycle and fix nitrogen, and because of the low nitrogen inputs, plant growth will be limited (Zedler and Langis 1991). Conversely, in systems with high nitrogen inputs, the low organic matter in created wetlands might limit the system's ability to process nitrogen through denitrification because of low carbon availability (Faulkner and Richardson 1991), and thus constrain water quality improvement values. Over time, we would expect the organic matter of soils of wetland projects to increase. However, because we as yet have no data on how long it will take organic soils to develop, enhancing the percentage of organic matter may be the best way to accelerate the development and facilitate the development of related functions.

Augmenting the substrate of wetland projects with SMS (Owen et al. 1989) from a donor wetland will make the substrate more similar to that of natural wetlands, and provide a possible source of appropriate wetland plant propagules. In addition, because organic soils have a higher capacity for water retention and an increased proportion of this water is available for plant growth, the probability of wetland vegetation establishment is increased. Organic soils also have higher cation exchange rates and consequently a higher buffering capacity than do mineral soils (Brady 1974). Because organic matter has a high capacity to complex or adsorb metals and organics, the amount of organic matter in the substrate can influence the wetland's potential for pollutant retention.

The contours of the project should be graded before the destruction of the natural marsh so that the SMS can be transferred directly. In any case, the SMS should not be stockpiled longer than 30 days because of possible oxidation of the soil, possible release of metals that may be toxic to seedlings, and possible loss of viability of some seeds (Brooks 1990). When transferring the SMS to the project, it should be spread over the substrate carefully, with minimal handling, overturning or trampling. If SMS is not available, there may be readily available sources of waste organic matter to augment mineral soils, such as municipal leaf/grass compost, composted livestock bedding and manure (although seeds of aggressive weedy species may be present), and food processing wastes.

Although the role of mulching or augmenting the organic matter content of soils is not yet clearly understood, we recommend augmenting the soils of projects to make the organic matter content more like that of natural wetlands. Further research will then provide insight as to whether or not augmentation accelerates the development of these projects.

Summary

Will changes in the design of wetland projects cause them to develop faster and become more like natural wetlands? Will they be a better "fit" in the landscape? Our interpretation of the results from field studies so far, indicates this may be true. We suggest that better wetlands can be designed by modeling projects on local natural wetlands and on what was learned from earlier projects. We contend that this will lead to ecologically based performance criteria for wetland restoration and creation that will, in turn, lead to better management and protection of the resource.

Looking to the future, we intend to continue building the knowledge base on wetland restoration and creation through the application, testing, and evaluation of the concepts presented in this document. The research to be implemented by EPA's WRP in the coming years will attempt to fill some of the gaps we have identified in the course of our studies to date. As stated earlier, there is a paucity of long-term data on the development of wetland projects. The projects we have described will soon be five years older. It will be a priority for us to repeat at least one of the three studies (i.e., Connecticut, Florida, and Oregon) to generate the next part of the performance curves. In this way we can further document the development of these freshwater wetland projects.

We have reported on the most common type of mitigation project nationally, a pond with a fringe of emergent marsh. Although they are very common, they are not the only type of wetlands being restored and created, or the only type being studied. Table 6-3 summarizes the findings from recent studies of groups of wetland projects. We are looking forward to applying our Approach to projects involving other wetland types to begin documenting their performance and to expand the scope of our Approach. Specifically, we will begin focusing on the restoration of riparian systems in the arid West in the near future.

We maintain that consideration of ecological setting is important to evaluate and understand the functions of natural wetlands and the performance of projects. Determining the effects of different land uses on wetland function will be a major theme of our upcoming research. Such information is necessary for both the protection of the wetland resource and the success of restoration and creation projects. With knowledge of the effects of surrounding land uses, appropriate management strategies can be employed to protect key wetlands, e.g., the use of buffers. In addition, knowing how present and projected development of an area will affect wetland function can influence decisions on how to prioritize sites so that projects maximize ecological benefits.

Fundamentally, as we plan and implement new studies we will continue to treat existing projects as experiments in progress and promote the idea that we all must

"...learn by going where we need to go..." (Roethke, 1961).

CITATION	LOCATION	WETLAND TYPES	# OF PROJECTS	FINDINGS
Crabtree et al. (1990)	14 States	saltwater, brackish, and freshwater	17	Success was related to adequacy of planning, design, implementation and follow-through.
Crewz (1990)	Manatee and Sarasota Counties, Florida	salt marsh, mangrove habitat and freshwater	11	Features of sites that did not meet agreed-upon criteria in the permit plan, or the permit criteria did not address habitat trade offs adequately.
Erwin (1991)	Florida	forested and non-forested freshwater wetlands	40	Only 4 projects met all stated permit goals. 16 of the failed or incomplete projects were correctable, but 6 could not succeed under any circumstances, and 14 projects required more study to determine the feasibility of corrective actions.
Florida Department of Environmental Regulation (1991)	Florida	freshwater herbaceous and forested wetlands, and tidal herbaceous and mangrove wetlands	119	A high rate of noncompliance was found. Only 4 of the 63 projects reviewed were in full compliance with permit conditions. The ecological success of sites built was 12% for freshwater systems and 45% for tidal systems.
Gwin and Kentula (1990)	Portland, Oregon	palustrine emergent and open water systems	11	None of the projects were constructed as permitted or planned. A cumulative loss of 29% of the area to be created occurred, and vegetation occurring on the projects consisted primarily of volunteer species.
Gwin et al. (1991)	Tampa, Florida	palustrine emergent and open water systems	9	Correlations between area required and area as-built could not be made for 6 of the created wetlands because of inadequate information in the project files. Vegetation occurring on the created wetlands consisted primarily of volunteer species.
Owen (1990)	Wisconsin	freshwater wetlands, mostly ponds	18	Eleven of the 18 sites resulted in a net loss of area. Four sites resulted in wetland types partially or entirely the same as those lost. Two have a good chance of becoming the type that was lost. Nine have incorrect physical conditions. Three were not constructed.
Roberts (1991)	Florida	coastal marshes	22	Vegetation characteristics were highly variable, but properly planned, constructed and maintained sites provided viable wildlife habitat.

Table 6-3.   A summary of the findings of recent studies of groups of wetland projects.

REFERENCES

Abbruzzese, B., A.B. Allen, S. Henderson, and M.E. Kentula. 1988. Selecting sites for comparison with created wetlands, p. 291-297. In C.D.A. Rubec and R.P. Overend (Comp.), Proceedings of Symposium '87—Wetlands/Peatlands. Environment Canada, Ottawa, Ontario, Canada.

Adamus, P.R. and K. Brandt. 1990. Impacts on Quality of Inland Wetlands of the United States: A Survey of Indicators, Techniques, and Applications of Community Level Biomonitoring Data. EPA/600/3-90/073. U.S. Environmental Protection Agency, Environmental Research Laboratory, Corvallis, OR.

Anderson, J.R., E.E. Hardy, J.T. Roach, and R.E. Witmer. 1976. A Land Use and Land Cover Classification System for Use with Remote Sensor Data. U.S. Department of Interior, Geological Survey, Professional Paper 964. Washington, DC.

Bedford, B.L. and E.M. Preston. 1988. Developing the scientific basis for assessing cumulative effects of wetland loss and degradation on landscape functions: Status, perspectives, and prospects. **Environmental Management** 12(5):751-772.

Berger, J.J. (Ed.) 1989. Environmental Restoration: Science and Strategies for Restoring the Earth. Island Press, Washington, DC.

Brady, N.C. 1974. The Nature and Properties of Soils. 8th ed. MacMillan Publishing Company, Inc., New York, NY.

Brooks, R.P. 1990. Wetland and water body restoration and creation associated with mining, p. 529-548. In J.A. Kusler and M.E. Kentula (Eds.), Wetland Creation and Restoration: The Status of the Science. Island Press, Washington DC.

Brooks, R.P. and R.M. Hughes. 1988. Guidelines for assessing the biotic communities of freshwater wetlands, p. 276-280. In J.A. Kusler, M.L. Quammen, and G. Brooks (Eds.), Proceedings of the National Wetland Symposium: Mitigation of Impacts and Losses. Association of State Wetland Managers, Berne, NY.

Brooks, R.P., E.D. Bellis, C.S. Keener, M.J. Croonquist, and D.E. Arnold. 1991. A methodology for biological monitoring of cumulative impacts on wetland, stream, and riparian components of watersheds, p. 387-398. In J.A. Kusler and

S. Daly (Eds.), Proceedings of the International Symposium: Wetlands and River Corridor Management. Association of State Wetland Managers, Berne, NY.

Brower, J.E. and J.H. Zar. 1984. Field and Laboratory Methods for General Ecology. 2nd ed. William C. Brown Co., Dubuque, IA.

Brown, M.T. 1991. Evaluating Constructed Wetlands Through Comparisons with Natural Wetlands. EPA/600/3-91/058. U.S. Environmental Protection Agency, Environmental Research Laboratory, Corvallis, OR.

Brown, M.T., J. Schaefer, and K. Brandt. 1990. Buffer Zones for Water, Wetlands, and Wildlife in East Central Florida. Publication Number 89-07 and Florida Agricultural Experiment Station Journal Series Number T-00061. Center for Wetlands, University of Florida, Gainesville, FL.

Cahoon, D.R. and R.E. Turner. 1989. Accretion and canal impacts in a rapidly subsiding wetlands. II. Feldspar marker horizon technique. **Estuaries** 12(4):260-268.

Cairns, J., Jr. (Ed.). 1988. Rehabilitating Damaged Ecosystems, Volumes I and II. CRC Press, Boca Raton, FL.

Chabreck, R.H. 1988. Coastal Marshes: Ecology and Wildlife Management. University of Minnesota Press, Minneapolis, MN.

Confer, S.R. 1990. Comparison of Created and Natural Freshwater Palustrine-Emergent Wetlands in Connecticut. M.A. Thesis, Department of Botany, Connecticut College, New London, CT.

Confer, S.R. and W.A. Niering. In press. Comparison of created and natural freshwater emergent wetlands in Connecticut. **Wetlands Ecology and Management.**

The Conservation Foundation. 1988. Protecting America's Wetlands: An Action Agenda. The Final Report of the National Wetlands Policy Forum. Washington, DC.

Cowardin, L.M., V. Carter, F.C. Golet, and E.T. LaRoe. 1979. Classification of Wetlands and Deepwater Habitats of the United States. FWS/OBS-79/31. U.S. Fish and Wildlife Service, Washington, DC.

Crabtree, A., E. Day, A. Garlo, and G. Stevens. 1990. Evaluation of Wetland Mitigation Measures, Final Report: Volume I. U.S. Department of Transportation, Federal Highway Administration Report Number FHWA-RD-90-083. Washington, DC.

Crewz, D.W. 1990. Habitat-Mitigation Evaluations for Manatee-Sarasota Counties, Mid-Project Summary: Projects 1-11. Report to ManaSota 88, Palmetto, FL.

Dahl, T.E. and C.E. Johnson. 1991. Status and Trends of Wetlands in the Conterminous United States, Mid-1970s to Mid-1980s. U.S. Department of the Interior, Fish and Wildlife Service, Washington, DC.

Devore, J. and R. Peck. 1986. Statistics: The Exploration and Analysis of Data. West Publishing Company, St. Paul, MN.

Ehnenfield, J.G. 1983. The effects of changes in land use on swamps of the New Jersey pine barrens. **Biological Conservation** 25:353-375.

Erwin, K.L. 1991. An Evaluation of Mitigation in the South Florida Water Management District. Volume I. South Florida Water Management District. West Palm Beach, FL.

Erwin, K.L. 1990. Freshwater marsh creation and restoration in the Southeast, p. 233-266. In J.A. Kusler and M.E. Kentula (Eds.), Wetland Creation and Restoration: The Status of the Science. Island Press, Washington, DC.

Erwin, K.L. 1988. A quantitative approach for assessing the character of freshwater marshes and swamps impacted by development in Florida, p. 295-297. In J.A. Kusler, M.L. Quammen, G. Brooks (Eds.), Proceedings of the National Wetland Symposium: Mitigation of Impacts and Losses. Association of State Wetland Managers, Berne, NY.

Faulkner, S.P. and C.J. Richardson. 1991. Physical and Chemical Characteristics of Freshwater Wetland Soils, p. 41-72. In D.A. Hammer (Ed.), Constructed Wetlands for Wastewater Treatment. Lewis Publishers, Inc, Chelsea, MI.

Federal Interagency Committee for Wetland Delineation. 1989. Federal Manual for Identifying and Delineating Jurisdictional Wetlands. U.S. Army Corps of Engineers, U.S. Environmental Protection Agency, U.S. Fish and Wildlife Service, and U.S. Department of Agriculture Soil Conservation Service, Cooperative Technical Publication. Washington, DC.

Florida Department of Environmental Regulation. 1991. Report on the Effectiveness of Permitted Regulation. Tallahassee, FL.

Frayer, W.E., D.D. Peters, and H.R. Pywell. 1989. Wetlands of the California Central Valley: Status and Trends, 1939 to mid-1980s. U.S. Fish and Wildlife Service, Region 1, Portland, OR.

Frenkel, R.E. and J.C. Morlan. 1990. Restoration of the Salmon River Salt Marshes: Retrospect and Prospect. Final Report to U.S. Environmental Protection Agency, Region 10, Seattle, WA.

Gallant, A.L., T.R. Whittier, D.P. Larsen, J.M. Omernik, and R.M. Hughes. 1989. Regionalization as a Tool for Managing Environmental Resources. EPA/600/3-89/060. U.S. Environmental Protection Agency, Washington, DC.

Garbisch, E.W., Jr. 1986. Highways and Wetlands: Compensating Wetland Losses. Contract Report DOT-FH-11-9442. Federal Highway Administration, Office of Implementation, McLean, VA.

Golet, F.C., A.J.K. Calhoun, W.R. DeRagon, D.J. Lowry, and A.J. Gold. In press. The Ecology of Red Maple Swamps in the Glaciated Northeastern United States: A Community Profile. U.S. Fish and Wildlife Service, Biological Report. Washington, DC.

Gosselink, J.G., L.C. Lee, and T.A. Muir (Eds.). 1990. Ecological Processes and Cumulative Impacts: Illustrated by Bottomland Hardwood Wetland Ecosystems. Lewis Publishers, Inc., Chelsea, MI.

Gwin, S.E. and M.E. Kentula. 1990. Evaluating Design and Verifying Compliance of Wetlands Created Under Section 404 of the Clean Water Act in Oregon. EPA/600/3-90/061. U.S. Environmental Protection Agency, Environmental Research Laboratory, Corvallis, OR.

Gwin, S.E., M.E. Kentula, and D.L. Frostholm, in conjunction with R.L. Tighe. 1991. Evaluating Design and Verifying Compliance of Created Wetlands in the Vicinity of Tampa, Florida. EPA/600/3-91/068. U.S. Environmental Protection Agency, Environmental Research Laboratory, Corvallis, OR.

Hammer, D.A. 1992. Creating Freshwater Wetlands. Lewis Publishers, Inc., Chelsea, MI.

Hammer, D.A. (Ed.). 1989. Constructed Wetlands for Wastewater Treatment: Municipal, Industrial, and Agricultural. Lewis Publishers, Inc., Chelsea, MI.

Hitchcock, C.L. and A. Cronquist. 1981. Flora of the Pacific Northwest—An Illustrated Manual. 5th ed. University of Washington Press, Seattle, WA.

Holland, C.C. and M.E. Kentula. In press. Impacts of Section 404 permits requiring compensatory mitigation on wetlands in California. **Wetlands Ecology and Management**.

Holland, C.C. and M.E. Kentula. 1991. The Permit Tracking System (PTS): A User's Manual. EPA/600/8-91/054. U.S. Environmental Protection Agency, Environmental Research Laboratory, Corvallis, OR.

Hollands, G.G. 1990. Regional analysis of the creation and restoration of kettle and pothole wetlands, p. 281-298. In J.A. Kusler and M.E. Kentula (Eds.), Wetland Creation and Restoration: The Status of the Science. Island Press, Washington, DC.

Hook, D.D., W.H. McKee Jr., H.K. Smith, J. Gregory, V.G. Burrell Jr., M.R. DeVoe, R.E. Sojka, S. Gilbert, R. Banks, L.H. Stolzy, C. Brooks, T.D. Matthews and T.H. Shear. 1988. The Ecology and Management of Wetlands. Timber Press, Portland, OR.

Horner, R.R. and K.J. Raedeke. 1989. Guide for Wetland Mitigation Projects Monitoring. Report Number WA-RD 195.1. Washington State Department of Transportation, Seattle, WA.

Hughes, H.G. and T.M. Bonnicksen (Eds.). 1990. Restoration '89: The New Management Challenge. Proceedings of the First Annual Meeting of the Society for Ecological Restoration, Oakland, CA.

Hughes, R.M., E. Rexstad, and C.E. Bond. 1987. The relationship of aquatic ecoregions, river basins, and physiographic provinces to the ichthyogeographic regions of Oregon. **Copia** 1987: 423-432.

Jordan, W.R., M.E. Gilpin, and J.D. Aber (Eds.). 1987. Restoration Ecology: A Synthetic Approach to Ecological Research. Cambridge University Press, Cambridge, UK.

Kadlec, R.H. 1988. Monitoring wetland responses, p. 114-120. In J. Zelazny and J.S. Feierabend (Eds.). Proceedings of Conference: Increasing Our Wetland Resources. National Wildlife Federation, Washington, DC.

Kantrud, H.A., G.L. Krapu, and G.A. Swanson. 1989. Prairie Basin Wetlands of the Dakotas: A Community Profile. U.S. Fish and Wildlife Service. Biological Report 85(7.28). Washington, DC.

Kentula, M.E., J.C. Sifneos, J.W. Good, M. Rylko, and K. Kunz. 1992. Trends and patterns in Section 404 permitting requiring compensatory mitigation in Oregon and Washington. **Environmental Management** 16:109-119.

King, D.M. 1991a. Wetland Creation and Restoration: An Integrated Framework for Evaluating Costs, Expected Results and Compensation Ratios. Prepared for U.S. Environmental Protection Agency, Office of Policy, Planning, and Evaluation, Washington, DC.

King, D.M. 1991b. Economics: Costing out restoration. **Restoration & Management Notes** 9(1):15-20.

Krebs, C.J. 1989. Ecological Methodology. Harper and Row, New York, NY.

Kruczynski, W.L. 1990. Options to be considered in preparation and evaluation of mitigation plans, p. 555-570. In J.A. Kusler and M.E. Kentula (Eds.), Wetland Creation and Restoration: The Status of the Science. Island Press, Washington, DC.

Kusler, J.A. and M.E. Kentula (Eds.). 1990a. Wetland Creation and Restoration: The Status of the Science. Island Press, Washington, DC.

Kusler, J.A. and M.E. Kentula. 1990b. Executive Summary, p. xvii-xxv. In J.A. Kusler and M.E. Kentula (Eds.), Wetland Creation and Restoration: The Status of the Science. Island Press, Washington, DC.

Kusler, J.A. and S. Daly (Eds.). 1989. Proceedings of an International Symposium: Wetlands and River Corridor Management. Association of State Wetland Managers, Berne, NY.

Langis, R., M. Zalejko, and J.B. Zedler. 1991. Nitrogen assessments in a constructed and a natural salt marsh of San Diego Bay, California. **Ecological Applications** 1:40-51.

Larsen, D.P., J.M. Omernik, R.M. Hughes, C.M. Rohm, T.R. Whittier, A.J. Kinney, A.L. Gallant, and D.R. Dudley. 1986. The correspondence between spatial patterns in fish assemblages in Ohio streams and aquatic ecoregions. **Environmental Management** 10:815-828.

Leibowitz, N.C., L. Squires, and J.P. Baker. 1991. Research Plan for Monitoring Wetland Ecosystems. EPA/600/3-01/010. U.S. Environmental Protection Agency, Environmental Research Laboratory, Corvallis,

Leibowitz, S.G., E.M. Preston, L.Y. Arnaut, N.E. Detenbeck, C.A. Hagley, M.E. Kentula, R.K. Olson, W.D. Sanville, and R.R. Sumner. 1992. Wetlands Research Plan FY92-96: An Integrated Risk-Based Approach. EPA/600/R-92/060. U.S. Environmental Protection Agency, Environmental Research Laboratory, Corvallis, OR.

Leopold, Aldo. 1966. Sand County Almanac. Oxford University Press. New York, NY.

Ludwig, J.A. and J.F. Reynolds. 1988. Statistical Ecology: A Primer on Methods and Computing. John Wiley and Sons, New York, NY.

Lugo, A.E., S. Brown, and M. Brinson. 1990. Forested Wetlands. Ecosystems of the World 15. Elsevier, Amsterdam, The Netherlands.

Lyons, J. 1989. Correspondence between the distribution of fish assemblages in Wisconsin streams and Omernik's ecoregions. **American Midland Naturalist** 122:163-182.

Marble, A.D. 1990. A Guide to Wetland Functional Design. Report Number FHWA-IP-90-010. Federal Highway Administration, McLean, VA.

Marsh, W.M. 1978. Environmental Analysis for Land Use and Site Planning. McGraw-Hill Book Company, New York, NY.

Majumdar, S.K., R.P. Brooks, F.J. Brenner, and R.W. Tiner, Jr. (Eds.). 1989. Wetlands Ecology and Conservation: Emphasis in Pennsylvania. The Pennsylvania Academy of Science, Philadelphia, PA.

McVoy G.R. 1988. Advantages of open water/emergent wetlands for mitigation and a holistic approach to banking, p. 289-290. In J.A. Kusler, M.L. Quammen, and G. Brooks (Eds.), Proceedings of the National Wetland Symposium: Mitigation of Impacts and Losses, Association of State Wetland Managers, Berne, NY.

Mitsch, W.J. and J.G. Gosselink. 1986. Wetlands. Van Nostrand Reinhold Company, Inc., New York, NY.

Mitsch, W.J., M. Straskraba, and S.E. Jorgensen (Eds.). 1988. Wetland Modelling. Elsevier Science Publishing Company, Inc., New York, NY.

Murkin, H.R. (Ed.). 1984. Marsh Ecology Research Program Long-term Monitoring Procedures Manual. Delta Waterfowl Research Station, Manitoba, Canada.

National Research Council, Committee on Restoration of Aquatic Ecosystems—Science, Technology, and Public Policy, Water Science and Technology Board, Commission on Geosciences, Environment, and Resources. 1992. Restoration of Aquatic Ecosystems: Science, Technology, and Public Policy. National Academy Press, Washington, DC.

Neter, J., W. Wasserman, and M.J. Kutner. 1990. Applied Linear Statistical Models: Regression, Analysis of Variance, and Experimental Designs. IRWIN, Homewood, IL.

Newling, C.J. and Landin, M.C. 1985. Long-term Monitoring of Habitat Development at Upland and Wetland Dredged Material Disposal Sites, 1974-1982. Technical Report D-85-5. Environmental Laboratory. U.S. Army Engineers Waterways Experiment Station, Vicksburg, MS.

Niering, W.A. 1991. Wetlands of North America. Thomasson-Grant, Charlottesville, VA.

Novitzki, R.P. 1989. Wetland hydrology, p. 46-64. In S.K. Majumbar, R.P. Brooks, F.J. Brenner, and R.W. Tiner, Jr. (Eds.), Wetlands Ecology and Conservation: Emphasis in Pennsylvania. The Pennsylvania Academy of Science, Philadelphia, PA.

Novitzki, R.P. 1982. Hydrology of Wisconsin Wetlands. Information Circular 40. U.S. Department of the Interior, Geological Survey and University of Wisconsin-Extension, Geological and Natural History Survey, Madison, WI.

O'Brien, A.L. 1986. Hydrology and the construction of a mitigating wetland, p. 83-200. In J.S. Larson and C. Neill (Eds.), Mitigating Freshwater Wetland

Alteration in the Glaciated Northeastern United States: An Assessment of the Science Base. Publication 87-1. Environmental Institute, University of Massachusetts, Amherst, MA.

Omernik, J.M. 1987. Ecoregions of the conterminous United States. **Annals of the Association of American Geographers** 77(1):118-125. (map scale 1:7,500,000)

Owen, C.R. 1990. Effectiveness of Compensatory Wetland Mitigation in Wisconsin. Technical Report to the Wisconsin Wetlands Association, The Lake Michigan Federation, The American Clean Water Project. University of Wisconsin, Madison, WI.

Owen, C.R., Q.J. Carpenter, and C.B. DeWitt. 1989. Evaluation of Three Wetland Restorations Associated with Highway Projects. Technical Report. Institute for Environmental Studies, University of Wisconsin, Madison, WI.

Pacific Estuarine Research Laboratory. 1990. A Manual for Assessing Restored and Natural Coastal Wetlands with Examples from Southern California. California Sea Grant Report Number T-CSGCP-021. LaJolla, CA.

Pielou, E.C. 1986. Assessing the diversity and composition of restored vegetation. **Canadian Journal of Botany** 64:1344-1348.

Quammen, M.L. 1986. Measuring the success of wetlands mitigation. **National Wetlands Newsletter** 8(5):6-8.

Reed, P.B., Jr. 1988. National List of Plant Species That Occur in Wetlands: Northwest (Region 9). U.S. Fish and Wildlife Service. Biological Report 88(26.9). Washington, DC. (This cite refers to the Pacific Northwest Volume. Other volumes are appropriate for specific areas).

Roberts, T.H. 1991. Habitat Value of Man-made Coastal Marshes in Florida. Technical Report WRP-RE-2. U.S. Army Engineers Waterways Experiment Station, Vicksburg, MS.

Roethke, Theodore. 1961. Words for the Wind. Indiana University Press, Bloomington, IN.

Rohm, C.M., J.W. Giese, and C.C. Bennett. 1987. Evaluation of an aquatic ecoregion classification of streams in Arkansas. **Freshwater Ecology** 4:127-140.

Schemnitz, S. 1980. Wildlife Management Techniques Manual. 4th ed. The Wildlife Society, Washington, DC.

Schneller-McDonald, K., L.S. Ischinger, and G.T. Auble. 1989. Wetland Creation and Restoration: Description and Summary of the Literature. U.S. Fish and Wildlife Service, Biological Report 89. Washington, DC.

Segelquist, C.A., W.L. Slauson, M.L. Scott, and G.T. Auble. 1990. Synthesis of Soil-Plant Correspondence Data From Twelve Wetland Studies Throughout the United States. U.S. Fish and Wildlilfe Service, Biological Report 90(19). Washington, DC.

Sharitz, R.R. and J.W. Gibbons (Eds.). 1989. Freshwater Wetlands and Wildlife. CONF-8603101, Department of Energy Symposium Series No. 61. U.S. Department of Energy, Office of Scientific and Technical Information, Oak Ridge, TN.

Sherman, A.D. 1991. Final Quality Assurance Report for the Connecticut Wetlands Study. EPA/600/3-91/030. U.S. Environmental Protection Agency, Environmental Research Laboratory, Corvallis, OR.

Sherman, A.D., S.E. Gwin, and M.E. Kentula, in conjunction with W.A. Niering. 1991. Quality Assurance Project Plan: Connecticut Wetlands Study. EPA/600/3-91/029. U.S. Environmental Protection Agency, Environmental Research Laboratory, Corvallis, OR.

Sifneos, J.C., E.W. Cake, Jr., and M.E. Kentula. In press(a). Effects of Section 404 permitting on freshwater wetlands in Louisiana, Alabama, and Mississippi. **Wetlands**.

Sifneos, J.C., M.E. Kentula, and P. Price. In press(b). Impacts of Section 404 permitting on freshwater wetlands in Texas and Arkansas. **Texas Journal of Science**.

Simenstad, C.A., C.D. Tanner, R.M. Thom, and L.D. Conquest. 1991. Estuarine Habitat Assessment Protocol. EPA/910/9-91-037. U.S. Environmental Protection Agency, Region 10, Office of Puget Sound, Seattle, WA.

Simon, B.D., L.J. Stoerzer, and R.W. Watson. 1988. Evaluating wetlands for flood storage, p. 104-109. In J.A. Kusler and G. Brooks, (Eds.), Proceedings of the National Wetland Symposium: Wetland Hydrology. Association of State Wetland Managers, Berne, NY.

Snedecor, G.W. and W.G. Cochran. 1980. Statistical Methods. The Iowa State University Press, Ames, IA.

Sokal, R.R. and F.J. Rohlf. 1981. Biometry. W.H. Freedman and Company, New York, NY.

Stark, N. 1972. Low Maintenance Vegetation: Wildland Shrubs, Their Biology and Utilization. General Technical Report INT-1. U.S. Department of Agriculture, Forest Service. Washington, DC.

Steward, A.N., L.J. Dennis, and H.M. Gilkey. 1963. Aquatic Plants of the Pacific Northwest. 2nd ed. Oregon State University Press, Corvallis, OR.

Thornburg, A. 1977. Use of vegetation for stabilization of shorelines of the Great Lakes, p. 39-53. In Proceedings of the Workshop on the Role of Vegetation in Stabilization of the Great Lakes Shoreline. Great Lakes Basin Commission, Ann Arbor, MI.

Tiner, R.W., Jr. 1984. Wetlands of the United States: Current Status and Recent Trends. U.S. Fish and Wildlife Service, National Wetland Inventory, Washington, DC.

Tiner, R.W., Jr. 1988. Field Guide to Nontidal Wetland Identification. Maryland Department of Natural Resources, Annapolis, MD and U.S. Fish and Wildlife Service, Newton Corner, MA. Cooperative publication.

U.S. Department of Agriculture, Soil Conservation Service. 1992. Field Handbook. Chapter 13: Wetland Restoration, Enhancement, and Creation. Washington, DC.

U.S. Fish and Wildlife Service. 1980. Habitat Evaluation Procedures. ESM 102. U.S. Department of the Interior, Fish and Wildlife Service, Division of Ecological Services, Washington, DC.

van der Valk, A.G. (Ed.). 1989. Northern Prairie Wetlands. Iowa State University Press, Ames, IA.

van der Valk, A.G. and C.B. Davis. 1976. The seed banks of prairie glacial marshes. **Canadian Journal of Botany** 54:1832-1838.

Verner, S.S. 1990. Handbook for Preparing Quality Assurance Project Plans for Environmental Measurements. Technical Resources, Inc., Rockville, MD.

Weinhold, C.E. and A.G. van der Valk. 1988. The impact of duration of drainage on the seed banks of northern prairie wetlands. **Canadian Journal Botany** 67:1878-1884.

Weller, M.W., G.W. Kaufmann, and P.A. Vohs. 1991. Evaluation of wetland development and waterbird response at Elk Creek Wildlife Management Area, Lake Mills, Iowa, 1961-1990. **Wetlands** 11(2):245-262.

Wentworth, T.R., G.P. Johnson, and R.L. Kologiski. 1988. Designation of wetlands by weighted averages of vegetation data: a preliminary evaluation. **Water Resources Bulletin** 24(2):389-396.

White, T.A., R. Lea, R.J. Haynes, W.L. Nutter, J.R. Nawrot, M.M. Brinson, and A.F. Clewell. 1990. Development and summary of MiST: a classification system for preproject mitigation sites and criteria for determining successful replication of forested wetlands, p. 323-335. In J. Skousen and J. Sencindiver (Eds.), Proceedings of the 1990 Mining and Reclamation Conference and Exhibition. American Society of Surface Mining and Reclamation, Charleston, WV.

Whittier, T.R., R.M. Hughes, and D.P. Larsen. 1988. Correspondence between ecoregions and spatial patterns in stream ecosystems in Oregon. **Canadian Journal of Fisheries and Aquatic Science** 45:1254-78.

Whittier, T.R., D.P. Larsen, R.M. Hughes, C.M. Rohm, A.L. Gallant, and J.M. Omernik. 1987. Ohio Stream Regionalization Project: A Compendium of Results. Freshwater Research Laboratory, Corvallis, OR.

Winter, T.C. 1981. Uncertainties in estimating the water balance of lakes. **Water Resources Bulletin.** 17(1): 82-115.

Zedler, J.B. and M.E. Kentula. 1986. Wetlands Research Plan. EPA/600/3-86/009. U.S. Environmental Protection Agency, Environmental Research Laboratory, Corvallis, OR.

Zedler, J.B. and R. Langis. 1991. Comparisons of constructed and natural salt marshes of San Diego Bay. **Restoration and Management Notes** 9(1):21-25.

INDEX

1990 Farm Bill, 29
Alabama, 12, 144
areas at risk, 13, 24
Arkansas, 5, 12, 17, 143, 144
as-built assessment, 44, 52, 56, 57, 61
as-built conditions, 44, 52, 59-61, 64
assessment, i, 12, 17, 43, 44, 52, 56, 57, 59-61, 66-69, 72, 143, 144
  procedures, 23, 37, 43, 44, 59, 60, 66, 68-70, 72, 80, 81, 96, 142, 145
base map, 56, 61
basin, 111, 113, 114, 116-118, 130, 140, 145
buffers, 116, 132
California, 12, 17, 26, 138-140, 143
characterization curves, 87, 93, 96, 107
Clean Water Act, 11, 74, 138
  Section 401, 15
  Section 404, xi, 12, 13, 15, 17, 19, 24, 35, 74, 112, 138-140, 144
comparability, 69, 70, 80
compensatory mitigation, 1, 13, 17, 26, 139, 140
compiling information, 13, 15, 29
compliance, 2, 17, 43, 44, 52, 72, 130, 138
comprehensive assessment, 17, 57, 60
confidence interval, 97, 102
Connecticut, i, 88, 105, 112, 117, 127, 131, 136, 144
construction plans, 44, 52, 127, 130
contouring, 114, 117
criteria, 2, 3, 8, 9, 36, 37, 44, 57, 59, 60, 63, 69, 72, 87, 98, 102, 104, 105, 108, 131, 146
Dakotas, 130, 140
data, i, xi, 3, 11-13, 15, 17, 19, 24, 26, 29, 43, 44, 52, 56, 57, 59-61, 64, 66, 68-76, 78-82, 87, 88, 92, 93, 96-98, 100, 102, 103-105, 107, 108, 112, 113, 129-131, 135, 137, 144, 146
  analysis, 3, 11, 13, 15, 26, 60, 68, 96, 98, 112, 120, 137, 139, 141, 142
  collection, 44, 52, 56, 57, 59, 60, 68, 69, 71, 74-76, 78-82, 87, 98
  entry, 15, 68, 98
  management system, 13, 15
  retrieval, 13
design, 3, 5, 8, 9, 24, 44, 52, 57, 59, 60, 70-72, 75, 77, 97, 111, 113, 114, 117, 127, 129-131, 138, 141
  guidelines, 3, 9, 72, 111, 120, 123, 135
dominance measure, 124
dominant species, 64, 123, 124
donor wetland, 122, 123, 130
ecological setting, 5, 33, 35, 37, 75, 132
ecoregion(s), 5, 17, 33, 35, 139, 141, 143, 146
emergent marsh, 29, 105, 112, 120, 125, 131
endemic, 125
erosion, 11, 122-124
evaluation, 2, 7, 12, 43, 44, 52, 59, 60, 63, 66, 68, 74, 76, 87, 88, 92, 105, 131, 137, 140, 143, 145, 146
exotics, 119, 125
fauna, 61, 66, 77, 107
field study
  Connecticut Study, 105, 127
  Florida Study, 29, 35, 92, 120

Oregon Study, 29, 33, 35-37, 52,
    92, 98, 103, 113, 114, 119,
    120, 130
flood detention, 112
Florida, i, 26, 29, 35, 88, 92, 105,
    107, 113, 117, 118, 120, 131,
    136, 137, 138, 143
freshwater nontidal wetlands, 3
groundwater, 117, 129
growing season, 66, 69, 71, 127
habitat, 19, 52, 56, 66, 71, 73, 97,
    112, 113, 137, 142-145
Habitat Evaluation Procedures
    (HEP), 66
herbaceous, 64, 100, 102, 103, 122,
    124, 125
homogenous, 29, 33, 97, 98, 100
hydrology, xi, 8, 44, 52, 61, 63, 71,
    76, 103, 113, 118, 120, 122, 127,
    129, 142, 144
hydroperiod, 118, 127, 129
hydrophytes, 8, 117
impacted wetlands, 12, 13, 15, 111,
    112
indicator(s), 7-8, 23, 43, 57, 63, 66,
    92, 93, 103, 135
inland wetlands, 135
invasive, 119, 125
jurisdictional wetland, 63, 93
land use, 5, 9, 35, 37, 52, 61, 88,
    135, 137, 141
landscape, 1, 2, 35, 60, 67, 68, 112,
    131, 135
long-term research, i, 59
Louisiana, 12, 17, 26, 144
maintenance, 123, 145
mapping, 56, 61, 74, 76
maps, 33, 36, 37, 52, 56, 61, 112
metals, 67, 131
Mississippi, 12, 144
mitigation, i, xi, 1-3, 7, 12, 13, 15,
    17, 23, 24, 26, 29, 33, 35, 72, 74,

75, 79, 113, 131, 135, 137, 139-
    143, 146
moisture gradient, 118
monitoring, 2, 3, 7, 8, 13, 17, 23,
    24, 26, 37, 43, 44, 52, 59-61, 63,
    66, 67, 71-75, 79, 82, 87, 92, 96,
    97, 108, 123, 135, 139-142
morphometry, 44, 61, 63
mulch, 100, 104, 122, 123, 131
National Wetlands Inventory, 26
nitrogen, 130, 140
nitrogen fixation, 130
nurseries, 124, 127
nutrients, 64, 67, 130
open water, 29, 36, 52, 92, 112,
    113, 127, 129, 141
opportunistic species, 125
Oregon, i, xi, 5, 12, 15, 17, 26, 29,
    33, 35-37, 52, 73-75, 88, 92, 98,
    103, 105, 107, 111-114, 117-120,
    127, 129-131, 138-140, 145, 146
organic, 63, 64, 88, 93, 96, 97, 100,
    107, 122, 130, 131
    matter, 63, 64, 88, 93, 96, 97,
        100, 107, 130, 131
    soils, 5, 8, 56, 63, 70, 76, 103,
        120, 122, 127, 130, 131, 135,
        137
outliers, 92, 100, 104
paired wetlands, 105
palustrine, 17, 36, 112, 117, 120,
    125, 136
    emergent marshes, 36, 120
    emergent wetland, 17, 119, 120
    forested wetland, 17
percent, 63, 64, 88, 92, 93, 96-98,
    100, 102, 104, 107, 124, 130
    of open water, 92, 112, 113, 141
    of species in common, 102
    organic matter, 63, 64, 88, 93, 96,
        97, 100, 107, 130, 131
performance, 2, 3, 5, 7-9, 23, 24,

35, 37, 43, 44, 52, 56, 57, 59, 60,
69, 72, 87, 88, 92, 93, 96-98,
100, 102-105, 108, 131, 132
curves, 5, 7-9, 23, 24, 87, 88, 93,
96, 104, 105, 107, 131
permanent sampling plots, 68
permit, xi, 2, 3, 12, 13, 15, 17, 19,
26, 37, 43, 44, 52, 60, 61, 66, 68,
72, 112, 120, 139
conditions, 5, 33, 35, 37, 43, 44,
52, 56, 59-61, 63, 64, 66-72,
74, 120, 122, 124, 129
record, 12, 13, 15, 26, 52, 56, 57,
61, 63, 67, 68, 76, 80, 112, 114
specifications, 17, 127
tracking system, xi, 13, 139
permitting, 11-13, 15, 17, 19, 24,
29, 44, 56, 59, 140, 144
activity, 11, 13, 17, 24, 26, 64, 72
agencies, 1, 11, 29, 73, 74, 77,
81, 87
assessment of the effects of, 12
cumulative impacts of, 11, 13
systems, 11, 15, 35, 61, 82, 92,
113, 130, 132
trends in, 11-13, 19, 26, 108
plant community, 64, 66, 124
composition, 63, 66, 104, 105,
118, 120, 123, 143
cover, 33, 64, 98, 100, 102-104,
120, 124, 135
planting lists, 118, 119
pond(s), 98, 105, 112-114, 116,
118, 129, 131
Portland, Oregon, xi, 17, 26, 29, 33,
35, 36, 98, 103, 104, 112, 138,
139
post-construction monitoring, 43
precision, 29, 36, 69, 71
progressive mean, 36
quality assurance objectives, 60
red maple swamps, 130, 138

reference sites, 3, 5, 35
region, 11, 35, 57, 71, 72, 74, 75,
112-114, 120, 124, 125, 138,
143, 144
regional flora, 125
regression, 96, 97, 142
regulation, 1, 11, 138
regulatory decisions, i, 1, 17
relative, 5, 9, 33, 35, 114, 120
abundance, 24, 63, 66, 68, 120, 124
elevations, 113, 114
representative sample, 26, 36
representativeness, 69, 70
restoration, 2, i, ii, xi, 2, 3, 5, 7, 9,
24, 35, 60, 69, 71, 92, 111, 114,
108, 120, 123, 127, 131, 132,
135, 137-140, 142, 144-146
revegetation, 120, 122
Rhode Island, 130
riparian system, 120
risk, 11, 13, 24, 26, 29, 33, 37, 92,
141
Rivers and Harbors Act, Section 10,
11
routine assessment, 57, 59, 60
sample, 23, 24, 26, 29, 35, 36, 59,
66, 70, 71, 88, 92, 97, 98, 100,
102, 104, 105, 114
sample size, 29, 70
stratified, 35, 70
sampling, 23, 24, 26, 36, 37, 60, 64,
66, 68-72, 80, 88, 100, 104, 123
design, 70
efficiency, 68, 69
protocols, 60, 66
strategy, 3, 17, 23, 37, 68-70, 72,
123
saturated, 63, 129
Seaside, Oregon, 75
sediment retention, 57, 112
seed bank, 123
seeds, 122, 124, 131

setting priorities, 23
Shrub/scrub wetland, 120
site selection, i, 5, 29, 37, 70
slope(s), 61, 63, 96-98, 112-114,
   116-118, 120
Society for Ecological Restoration,
   139
soil, 29, 63, 64, 71, 93, 97, 107,
   116, 117, 122, 125, 130, 131,
   137, 144, 145
   augmentation, 122, 131
   gleyed, 63
   hydric, 8, 56, 63, 71
   microflora, 122
   mineral, 131
   mottles, 63
   organic matter, 63, 64, 88, 93, 96,
      97, 100, 107, 130, 131
   pore water, 122
   saturation, 63
   stabilization, 125, 145
species, xi, 7, 19, 52, 60, 64, 66, 68,
   71, 73, 88, 98, 100, 102-105,
   118, 119, 120, 122-125, 127,
   131, 143
   composition, 63, 66, 104, 105,
      118, 120, 123, 143
   diversity, 7, 8, 43, 88, 100, 104,
      105, 143
   diversity index, 88
   plant diversity, 100, 105
sprigs, 124
staff gauge, 129
standard operating procedures, 60
state-wide standardization, 13
statistical tests, 96, 108
   F-test, 97
   hypothesis tests, 97
   Levine's test, 97
   statistical analyses, 96
   statistics, 137
Student's T-test, 96, 98

structural characteristics, 112, 116,
   127
study area, 26, 33, 36
substrate(s), 44, 52, 59, 61, 63, 64,
   88, 96, 97, 100, 122, 125, 127,
   129-131
success, 2, 3, 71-74, 76-80, 87, 108,
   123, 127, 132, 143
successional species, 125
surface water, 63, 105
Tampa, Florida, 138
Texas, 12, 17, 19, 144
timing of sampling, 26
topographical profiles, 114
training, 56, 63, 69, 70, 76, 79-81
transects, 61, 70, 71, 98, 114
transitional area, 113, 116-118
trends in permitting, 11, 13, 19
upland, 71, 103, 114, 116-118, 142
variability, 3, 29, 52, 57, 67, 70, 92,
   97, 100, 118
variance, 142
variable, 8, 63, 69, 70, 88, 93, 105
vegetation, 5, 44, 52, 56, 61, 64, 66,
   69-71, 76, 98, 100, 102-104, 107,
   111, 112-114, 118-120, 122-125,
   127, 130, 143, 145, 146
   communities, 56, 59, 63, 64, 66,
      69, 75, 98, 118-120, 123, 135
   cover, 98, 103
   emergent vegetation, 98, 118
   herbaceous vegetation, 100, 103
   percent cover, 98, 100, 104, 124
   stratum, 70, 124
   zonation, 118
volunteer species, 119, 120
Washington, 12, 17, 73, 124, 135-
   140, 142-145
water, i, 11, 29, 36, 43, 52, 57, 61,
   63, 67, 68, 70, 71, 74, 76, 80, 92,
   93, 97, 105, 107, 111-114, 116,
   117, 122, 125, 127, 129, 130,

135-138, 141-143, 146
budgets, 129
control structures, 52
depth, 59, 63, 68, 113, 117, 127, 130
level, i, 1, 2, 5, 7-9, 36, 44, 60, 63, 68, 69, 75, 79, 82, 88, 92, 93, 97, 98, 105, 107, 116, 122, 129, 130, 135
level recorders, 129
quality, 5, 7, 12, 13, 35, 36, 60, 67, 69, 72, 73, 75, 76, 79, 80, 82, 97, 130, 135, 144, 145
retention, 57, 112, 130, 131
source, 69, 96, 116, 117, 122, 130
table, xi, 12, 13, 15, 29, 36, 44, 61, 63, 68, 107, 113, 116, 117, 125, 127, 132
waterfowl, 29, 113, 142
watershed, 5, 61, 93
weedy species, 131
weighted average, 92, 103
wetland management, 1, 7, 11, 24, 71, 108
management decisions, 1, 3, 11, 12, 15, 17, 19, 82
managers, i, 3, 8, 13, 17, 59, 74, 82, 87, 92, 96, 108, 135-137, 140, 142, 144
Willamette Valley, Oregon, xi, 112
Wisconsin, 5, 111, 113, 141-143
woody cuttings, 124
woody species, 122, 125